# CONTEMPORARY MATHEMATICS

## Titles in This Series

**Volume**

1 **Markov random fields and their applications**, Ross Kindermann and J. Laurie Snell

2 **Proceedings of the conference on integration, topology, and geometry in linear spaces**, William H. Graves, Editor

3 **The closed graph and P-closed graph properties in general topology**, T. R. Hamlett and L. L. Herrington

4 **Problems of elastic stability and vibrations**, Vadim Komkov, Editor

5 **Rational constructions of modules for simple Lie algebras**, George B. Seligman

6 **Umbral calculus and Hopf algebras**, Robert Morris, Editor

7 **Complex contour integral representation of cardinal spline functions**, Walter Schempp

8 **Ordered fields and real algebraic geometry**, D. W. Dubois and T. Recio, Editors

9 **Papers in algebra, analysis and statistics**, R. Lidl, Editor

10 **Operator algebras and K-theory**, Ronald G. Douglas and Claude Schochet, Editors

11 **Plane ellipticity and related problems**, Robert P. Gilbert, Editor

12 **Symposium on algebraic topology in honor of José Adem**, Samuel Gitler, Editor

13 **Algebraists' homage: Papers in ring theory and related topics**, S. A. Amitsur, D. J. Saltman, and G. B. Seligman, Editors

14 **Lectures on Nielsen fixed point theory**, Boju Jiang

15 **Advanced analytic number theory. Part I: Ramification theoretic methods**, Carlos J. Moreno

16 **Complex representations of GL(2, K) for finite fields K**, Ilya Piatetski-Shapiro

17 **Nonlinear partial differential equations**, Joel A. Smoller, Editor

18 **Fixed points and nonexpansive mappings**, Robert C. Sine, Editor

19 **Proceedings of the Northwestern homotopy theory conference**, Haynes R. Miller and Stewart B. Priddy, Editors

20 **Low dimensional topology**, Samuel J. Lomonaco, Jr., Editor

21 **Topological methods in nonlinear functional analysis**, S. P. Singh, S. Thomeier, and B. Watson, Editors

22 **Factorizations of $b^n \pm 1$, $b = 2, 3, 5, 6, 7, 10, 11, 12$ up to high powers**, John Brillhart, D. H. Lehmer, J. L. Selfridge, Bryant Tuckerman, and S. S. Wagstaff, Jr.

23 **Chapter 9 of Ramanujan's second notebook—Infinite series identities, transformations, and evaluations**, Bruce C. Berndt and Padmini T. Joshi

24 **Central extensions, Galois groups, and ideal class groups of number fields**, A. Fröhlich

25 **Value distribution theory and its applications**, Chung-Chun Yang, Editor

26 **Conference in modern analysis and probability**, Richard Beals, Anatole Beck, Alexandra Bellow, and Arshag Hajian, Editors

## Titles in This Series

**Volume**


27 **Microlocal analysis,** M. Salah Baouendi, Richard Beals, and Linda Preiss Rothschild, Editors

28 **Fluids and plasmas: geometry and dynamics,** Jerrold E. Marsden, Editor

29 **Automated theorem proving,** W. W. Bledsoe and Donald Loveland, Editors

30 **Mathematical applications of category theory,** J. W. Gray, Editor

31 **Axiomatic set theory,** James E. Baumgartner, Donald A. Martin, and Saharon Shelah, Editors

32 **Proceedings of the conference on Banach algebras and several complex variables,** F. Greenleaf and D. Gulick, Editors

33 **Contributions to group theory,** Kenneth I. Appel, John G. Ratcliffe, and Paul E. Schupp, Editors

34 **Combinatorics and algebra,** Curtis Greene, Editor

35 **Four-manifold theory,** Cameron Gordon and Robion Kirby, Editors

36 **Group actions on manifolds,** Reinhard Schultz, Editor

37 **Conference on algebraic topology in honor of Peter Hilton,** Renzo Piccinini and Denis Sjerve, Editors

38 **Topics in complex analysis,** Dorothy Browne Shaffer, Editor

39 **Errett Bishop: Reflections on him and his research,** Murray Rosenblatt, Editor

40 **Integral bases for affine Lie algebras and their universal enveloping algebras,** David Mitzman

41 **Particle systems, random media and large deviations,** Richard Durrett, Editor

42 **Classical real analysis,** Daniel Waterman, Editor

43 **Group actions on rings,** Susan Montgomery, Editor

44 **Combinatorial methods in topology and algebraic geometry,** John R. Harper and Richard Mandelbaum, Editors

45 **Finite groups–coming of age,** John McKay, Editor

46 **Structure of the standard modules for the affine Lie algebra $A_1^{(1)}$,** James Lepowsky and Mirko Primc

47 **Linear algebra and its role in systems theory,** Richard A. Brualdi, David H. Carlson, Biswa Nath Datta, Charles R. Johnson, and Robert J. Plemmons, Editors

48 **Analytic functions of one complex variable,** Chung-chun Yang and Chi-tai Chuang, Editors

49 **Complex differential geometry and nonlinear differential equations,** Yum-Tong Siu, Editor

50 **Random matrices and their applications,** Joel E. Cohen, Harry Kesten, and Charles M. Newman, Editors

# CONTEMPORARY MATHEMATICS

Volume 48

# Analytic Functions of One Complex Variable

Chung-chun Yang and
Chi-tai Chuang, Editors

AMERICAN MATHEMATICAL SOCIETY
Providence · Rhode Island

1980 *Mathematics Subject Classification.* Primary 30-02

**Library of Congress Cataloging-in-Publication Data**
Main entry under title:
Analytic functions of one complex variable.
(Contemporary mathematics, ISSN 0271-4232; v. 48)
Bibliography: p.
1. Functions of complex variables. 2. Analytic functions. I. Yang, Chung-Chun, 1942– . II. Chuang, Chi-tai. III. Series: Contemporary mathematics (American Mathematical Society); v. 48.
QA331.A6547 1985 515.9 85-23004
ISBN 0-8218-5050-4 (alk. paper)

CONTENTS

PREFACE vii

CONTRIBUTORS TO THIS VOLUME ix

ON THE INVERSION OF A LINEAR DIFFERENTIAL POLYNOMIAL 1
Chi-tai Chuang

DISTRIBUTION OF THE VALUES OF MEROMORPHIC FUNCTIONS 21
Chi-tai Chuang and Lo Yang

A SIMPLE PROOF OF A THEOREM OF FATOU ON THE ITERATION AND FIX-POINTS OF TRANSCENDENTAL ENTIRE FUNCTIONS 65
Chi-tai Chuang

ALGEBROID FUNCTIONS AND THEIR APPLICATIONS TO ORDINARY DIFFERENTIAL EQUATIONS 71
Yu-zan He

SUMMARY OF RECENT RESEARCH ACCOMPLISHMENTS IN VALUE DISTRIBUTION THEORY AT EAST CHINA NORMAL UNIVERSITY 85
Rui-fu Lee, Chong-ji Dai, and Guo-dong Song

UNIVALENT FUNCTIONS 101
Shu-qi Liu and Chi-tai Chuang

QUASICOMFORMAL MAPPINGS 129
Cheng-qi He and Zhong Li

RIEMANN SURFACES 151
Ming-yong Zhang

APPROXIMATION AND INTERPOLATION IN THE COMPLEX DOMAIN 157
Xi-chang Shen

DIRICHLET SERIES 201
Chia-yung Yu

APPLICATION OF COMPLEX ANALYSIS TO NONLINEAR ELLIPTIC SYSTEMS OF PARTIAL DIFFERENTIAL EQUATIONS 217
Guo-chun Wen

RIEMANN ZETA-FUNCTION 235
Nan-yue Zhang and Shun-yan Zhang

APPLICATION OF THE THEORY OF FUNCTIONS OF A COMPLEX VARIABLE TO CLESTIAL MECHANICS 243
Zhao-hua Yi

COMPLEX VARIABLE METHODS IN ELASTICITY 247
Jian-ke Lu

## PREFACE

In the past, Chinese scholars had won brilliant accomplishments with inventions and creations in the evolution of science and technology, and particularly, made significant contributions in earlier development of mathematics.

Unfortunately, for about two decades, before and during the disruptive cultural revolution, the Chinese scientific community as a whole had been alienated from the western world and publication of works by Chinese scientists was very much discouraged.

Now, since the four modernization plans, China has reopened her doors and has rapidly taken the path to scientific and cultural interactions with the rest of the world

As mathematicians, we have planned to edit a series of mathematical monographs in various fields of mathematics: Complex Analysis, Harmonic Analysis, Functional Analysis, Differential and Integral Equations, Algebra, Topology and Differential Geometry, Number Theory, Numerical Analysis, and Applied Mathematics.

It is hoped that these collections will not only reflect the achievements and contributions of Chinese Mathematicians in the development of contemporary mathematics, but, more importantly, provide a broad view of the scope of and aid greatly those interested in keeping abreast of developments in mathematical research in modern China for the past three decades. Furthermore, we hope the publication of these volumes will further promote the scientific exchanges between West and East, and become a trigger for similar projects in other scientific disciplines.

Thus, the nature of the articles solicited are mostly expository and survey types and written by contemporary distinguished Chinese mathematicians in their respective fields. The theme presented in this volume is focused on complex analysis of one variable, which is a well developed field in China.

In presenting this first volume to the reader, we wish to thank the American Mathematical Society for their support and endorsement in such an endeavor; which also made this difficult task an enjoyable and rewarding experience.

Finally, allow us in the name of the Editors to express our heartfelt thanks and appreciation to all the contributors who have kindly prepared and sent us their articles toward the completion of this book.

Chung-chun Yang 楊重駿

Chi-tai Chuang 庄圻泰

# CONTRIBUTORS TO THIS VOLUME

Chi-tai Chuang 庄圻泰
The department of Mathematics, Peking University, Beijing, China

Chong-ji Dai 戴崇基
The Department of Mathematics, East China Normal University
Shanghai, China

Cheng-qi He 何成奇
The Department of Mathematics, Fudan University, Shanghai, China

Yu-zan He 何育赞
Institute of Mathematics, Academia Sinica, Beijing, China

Rui-fu Lee 李锐夫
The Department of Mathematics, East China Normal University
Shanghai, China

Zhong Li 李忠
The Department of Mathematics, Peking University, Beijing, China

Shu-qi Liu 刘书琴
The Department of Mathematics, Xian University, Xian, China

Jian-ke Lu 路见可
The Department of Mathematics, Wuhan University, Wuhan, China

Xi-chang Shen 沈燮昌
The Department of Mathematics, Peking University, Beijing, China

Guo-dong Song 宋国栋
The Department of Mathematics, East China Normal University
Shanghai, China

Guo-chun Wen 闻国椿
The Department of Mathematics, Peking University, Beijing, China

Lo Yang (or Le Yang) 杨乐
Institute of Mathematics, Academia Sinica, Beijing, China

Zhao-hua Yi 易照华
The Department of Astronomy, Nanjing University, Nanjing, China

Chia-yung Yu 余家荣
The Department of Mathematics, Wuhan University, Wuhan, China

Ming-yong Zhang 张鸣镛
The Department of Mathematics, Xiamen University, Xiamen, China

Nan-yue Zhang 张南岳
The Department of Mathematics, Peking University, Beijing, China

Shun-yan Zhang 张顺燕
The Department of Mathematics, Peking University, Beijing, China

Contemporary Mathematics
Volume 48, 1985

# ON THE INVERSION OF A LINEAR DIFFERENTIAL POLYNOMIAL

Chi-tai Chuang

In this paper meromorphic functions are always functions which are meromorphic for $|z| < +\infty$. For a meromorphic function $f(z)$, we shall use the associated notations $m(r,f)$, $N(r,f)$, $T(r,f)$ introduced by Nevanlinna [1].

Consider $p$ linearly independent meromorphic functions $\psi_k(z)(k=1,2,\ldots,p;p \geq 1)$ and a meromorphic function $f(z)$. Denote resprectively by $A_0 = \Delta(\psi_1,\psi_2,\ldots,\psi_p)$ and $\Delta(f,\psi_1,\psi_2,\ldots,\psi_p)$ the Wronskian determinant of $\psi_k(z)(k = 1,2,\ldots,p)$ and that of $f$, $\psi_k(k = 1,2,\ldots,p)$ and set [2]

$$L(f) = \frac{(-1)^p}{A_0}\Delta(f,\psi_1,\psi_2,\ldots,\psi_p) = f^{(p)} + \frac{A_1}{A_0}f^{(p-1)} + \ldots + \frac{A_{p-1}}{A_0}f' + \frac{A_p}{A_0}f. \tag{1}$$

We propose to express inversely the meromorphic function $f(z)$ in terms of the meromorphic function

$$F = L(f) \tag{2}$$

and prove the formula

$$f(z) = \sum_{k=1}^{p} C_k(z)\psi_k(z), \tag{3}$$

where $C_k(z)(k = 1,2,\ldots,p)$ are mermorphic functions satisfying the identities

$$C_k' = (-1)^{p+k}\frac{\Phi_k}{A_0}F \ (k = 1,2,\ldots,p) \tag{4}$$

and $\Phi_k$ is the determinant obtained by omitting the $p$th row and the $k$-th column of the matrix

$$\begin{pmatrix} \psi_1 & \psi_2 & \cdots & \psi_p \\ \psi_1' & \psi_2' & \cdots & \psi_p' \\ \cdots & \cdots & \cdots & \cdots \\ \psi_1^{(p-1)} & \psi_2^{(p-1)} & \cdots & \psi_p^{(p-1)} \end{pmatrix}. \tag{5}$$


0271-4132/85 $1.00 + $.25 per page

For this purpose, we proceed as follows: We consider the differential equation

$$L(w) = F, \tag{6}$$

where $w$ is the unknown function and $F$ is the meromorphic function defined by (2), and we try to find a solution of the equation (6) of the form

$$w = \sum_{k=1}^{p} C_k(z)\psi_k(z) \tag{7}$$

by a method of Lagrange called the method of variation of constants [3]. Accordingly we set

$$\psi_1^{(j)}C_1' + \psi_2^{(j)}C_2' + \ldots + \psi_p^{(j)}C_p' = 0 \quad (j = 0,1,\ldots,p-2). \tag{8}$$

Then

$$w^{(j)} = C_1\psi_1^{(j)} + C_2\psi_2^{(j)} + \ldots + C_p\psi_p^{(j)} \quad (j = 1,2,\ldots,p-1) \tag{9}$$

and

$$w^{(p)} = C_1\psi_1^{(p)} + C_2\psi_2^{(p)} + \ldots + C_p\psi_p^{(p)} + \psi_1^{(p-1)}C_1' + \psi_2^{(p-1)}C_2' + \ldots + \psi_p^{(p-1)}C_p'. \tag{10}$$

Substituting (7), (9), and (10) in (6), we get

$$L(w) = \sum_{k=1}^{p} C_k L(\psi_k) + \psi_1^{(p-1)}C_1' + \psi_2^{(p-1)}C_2' + \ldots + \psi_p^{(p-1)}C_p'$$

$$= \psi_1^{(p-1)}C_1' + \psi_2^{(p-1)}C_2' + \ldots + \psi_p^{(p-1)}C_p'.$$

We set

$$\psi_1^{(p-1)}C_1' + \psi_2^{(p-1)}C_2' + \ldots + \psi_p^{(p-1)}C_p' = F. \tag{11}$$

From (8) and (11) we can solve for $C_k'(k = 1,2,\ldots,p)$ and get (4).

Now suppose that we can find meromorphic functions $C_k(k = 1,2,\ldots,p)$ satisfying (4), then the meromorphic function $w$ defined by (7) is a solution of the equation (6) which may be written in the form

$$L(f - w) = 0.$$

Consequently

$$f - w = \sum_{k=1}^{p} b_k\psi_k, \tag{12}$$

where $b_k(k = 1,2,\ldots,p)$ are constants. From (7) and (12), we have

$$f = \sum_{k=1}^{p} (C_k + b_k)\psi_k.$$

Since $C_k + b_k (k = 1,2,...,p)$ are also meromorphic functions satisfying (4), the formula (3) follows.

Thus to prove the formula (3), it is sufficient to show that we can find meromorphic functions $C_k (k = 1,2,...,p)$ satisfying (4).

For this purpose, denote the right memebers of (4) by

$$g_k = (-1)^{p+k} \frac{\Phi_k}{A_0} F (k = 1,2,...,p) \tag{13}$$

and write (4) in the form

$$C'_k = g_k \quad (k = 1,2,...,p) \tag{14}$$

Let $z_0$ be a point which is a pole of at least one of the functions $g_k$ $(k = 1,2,...,p)$. Among the functions $g_k (k = 1,2,...,p)$, let $g_{k_j}$ $(j = 1, 2,...,q)$ be all those admitting $z_0$ as a pole. Consider a circle $\gamma : |z-z_0| < \delta$ such that in the domain $d : 0 < |z-z_0| < \delta$ the functions $\psi_k (k = 1,2,...,p)$, $1/A_0$ and $f$ are all holomorphic. Then $g_k (k = 1,2,..,p)$ are holomorphic in $d$. In $d$ we have

$$g_{k_j}(z) = p_j(z) + \varphi_j(z), \ p_j(z) = \frac{d_j}{z-z_0} + ... + \frac{p_j}{(z-z_0)^{m_j}}$$

$$(j = 1,2,...,q), \tag{15}$$

where $p_j(z)$ is the principal part of $g_{k_j}(z)$ at $z_0$ and $\varphi_j(z)$ is holomorphic in $\gamma$. We are going to show that

$$d_j = 0 \quad (j = 1,2,...,q). \tag{16}$$

In fact consider the domain

$$d_1 : 0 < |z-z_0| < \delta, \ 0 < \arg(z-z_0) < 2\pi \tag{17}$$

and the function

$$\log(z-z_0) = \log |z-z_0| + i\arg(z-z_0)$$

holomorphic in $d_1$, where $\arg(z-z_0)$ is defined by (17). We have

$$\{\log(z-z_0)\}' = \frac{1}{z-z_0} .$$

From (15), we can find a function $\lambda_j(z)$ holomorphic in $d$ such that

$$\frac{d}{dz} \{d_j \log(z-z_0) + \lambda_j(z)\} = g_{k_j}(z) \quad \text{in } d_1$$

$$(j = 1,2,...,q).$$

Now for the sake of definitess, suppose that $q = 3$ and $k_j = j$ $(j = 1,2,3)$, and set

$$\Gamma_j(z) = d_j \log(z-z_0) + \lambda_j(z) \quad (j = 1,2,3).$$

Then $\Gamma_j(z)$ $(j = 1,2,3)$ are holomorphic in $d_1$ and

$$\Gamma_j'(z) = g_j(z) \quad \text{in } d_1 \quad (j = 1,2,3).$$

On the other hand $g_j(z)$ $(j = 4,5,\ldots,p)$ being holomorphic in $\gamma$, we can find functions $G_j(z)$ $(j = 4,5,\ldots,p)$ holomorphic in $\gamma$ such that

$$G_j'(z) = g_j(z) \quad \text{in } \gamma \quad (j = 4,5,\ldots,p).$$

Consequently the function

$$w = \sum_{j=1}^{3} \Gamma_j(z)\psi_j(z) + \sum_{j=4}^{p} G_j(z)\psi_j(z)$$

satisfies the equation (6) in $d_1$, and we conclude as for (12) that

$$w - f = \sum_{k=1}^{p} b_k'\psi_k$$

in $d_1$, where $b_k'$ $(k = 1,2,\ldots,p)$ are constants. So in $d_1$ we have

$$\{d_1\psi_1(z) + d_2\psi_2(z) + d_3\psi_3(z)\} \log(z-z_0) = \Psi(z), \tag{18}$$

where $\Psi(z)$ is holomorphic in $d$. (18) implies that $d_j = 0$ $(j = 1,2,3)$. In fact, if this is untrue, then, since $\psi_k$ $(k = 1,2.,,,p)$ are linearly independent, $\sum_{j=1}^{3} d_j\psi_j(z)$ is not identically equal to zero in $d$. There is then a point $z' = z_0+h$ $(0 < h < \delta)$ such that

$$\zeta = \sum_{j=1}^{3} d_j\psi_j(z') \neq 0.$$

Now let $z \to z'$ in keeping first $\operatorname{Im} z > \operatorname{Im} z_0$ and then $\operatorname{Im} z < \operatorname{Im} z_0$, we get respectively from (18)

$$\zeta \log|z'-z_0| = \Psi(z'), \quad \zeta(\log|z' - z_0| + 2\pi i) = \psi(z')$$

and then $\zeta 2\pi i = 0$ which is impossible.

k being fixed, we are going to show that we can find a meromorphic function $C_k$ satisfying (14). We first suppose that the function $g_k(z)$ has an infinite number of poles $z_j$ $(j = 0,1,2,\ldots)$ such that

$$z_0 = 0, \; 0 < |z_1| \leq |z_2| \leq \ldots \leq |z_j| \leq |z_{j+1}| \leq \ldots$$

$$\lim_{j\to+\infty} |z_j| = +\infty.$$

Consider the series

$$u(z) = U_0(z) + \sum_{j=1}^{\infty} \{U_j(z) - P_j(z)\}, \tag{19}$$

where $U_j(z)(j \geq 0)$ is the principal part of $g_k(z)$ at $z_j$ and $P_j(z)(j \geq 1)$ is a polynomial of $z$ such that

$$|U_j(z) - P_j(z)| < \frac{1}{2^j} \quad \text{for} \quad |z| \leq \frac{1}{2}|z_j|. \tag{20}$$

It is easy to see that the series (19) is locally uniformly convergent in the domain $D = \mathbb{C} - \{z_j(j = 0,1,2,\ldots)\}$ and defines a meromorphic function $u(z)$ whose poles are the points $z_j(j = 0,1,2,\ldots)$ with the corresponding principal part $U_j(z)(j = 0,1,2,\ldots)$. Hence we have

$$g_k(z) = E(z) + u(z),$$

where $E(z)$ is an entire function. Consequently, it is sufficient to show that we can find a meromorphic function $v(z)$ such that

$$v'(z) = u(z). \tag{21}$$

Here the result (16) obtained above is essential. By this result, we can write

$$\begin{aligned} U_j(z) &= \sum_{n=2}^{h_j} \frac{a_{jn}}{(z-z_j)^n} \quad (j = 0,1,2,\ldots) \\ P_j(z) &= \sum_{n=0}^{k_j} b_{jn} z^n \quad (j = 1,2,\ldots). \end{aligned} \tag{22}$$

Let

$$\begin{aligned} v_0(z) &= \sum_{n=2}^{h_0} \frac{-a_{0n}}{n-1} \frac{1}{z^{n-1}} \\ v_j(z) &= \sum_{n=2}^{h_j} \frac{-a_{jn}}{n-1} \frac{1}{(z-z_j)^{n-1}} - \sum_{n=0}^{k_j} b_{jn} \frac{z^{n+1}}{n+1} \quad (j = 1,2,\ldots). \end{aligned} \tag{23}$$

and let $\zeta_0$ be a point of the domain $D$. We are going to show that the series

$$v(z) = \sum_{j=0}^{\infty} \{v_j(z) - v_j(\zeta_0)\} \tag{24}$$

defines a meromorphic function $v(z)$ satisfying (21). In fact, consider first a circle $\bar{\gamma}: |z-\zeta_0| \leq r$ such that $\bar{\gamma} \subset D$. Let $j_0$ be a positive integer such that $\bar{\gamma} \subset (|z| < \frac{1}{2}|z_{j_0}|)$. Then, from (20), for $j \geq j_0$, we have

$$|U_j(z) - P_j(z)| < \frac{1}{2^j} \quad \text{in } \bar{\gamma}.$$

Hence we can find a convergent series of positive terms $\sum_{j=0}^{\infty} M_j$ such that in $\bar{\gamma}$,

$$|U_0(z)| \leq M_0, \ |U_j(z) - P_j(z)| \leq M_j \ (j = 1,2,\ldots).$$

Now consider the circle $\gamma: |z-\zeta_0| < r$. In $\gamma$, the functions $U_0(z)$, $U_j(z) - P_j(z)$ $(j=1,2,\ldots)$ and $v_j(z) - v_j(\zeta_0)$ $(j=0,1,2,\ldots)$ are holomorphic and satisfy the relations

$$\{v_0(z) - v_0(\zeta_0)\}' = U_0(z), \ \{v_j(z) - v_j(\zeta_0)\}' = U_j(z) - P_j(z)$$

$$(j = 1,2,\ldots).$$

Hence in $\gamma$, we have

$$v_0(z) - v_0(\zeta_0) = \int_{\zeta_0}^{z} U_0(z)dz, \ v_j(z) - v_j(\zeta_0) = \int_{\zeta_0}^{z} \{U_j(z) - P_j(z)\}dz$$

$$(j = 1,2,\ldots),$$

$$|v_0(z) - v_0(\zeta_0)| \leq rM_0, \ |v_j(z) - v_j(\zeta_0)| \leq rM_j$$

$$(j = 1,2,\ldots).$$

Consequently the series (24) is absolutely and uniformly convergent in $\gamma$.

In general, we see in the same manner that if $j'$ is a non-negative integer and $\zeta'$ a point such that $\zeta' \neq z_j$ for $j \geq j'$ and if $\gamma': |z-\zeta'| \leq r'$ is a circle which contains none of the points $z_j$ $(j \geq j')$, then in the circle $\gamma': |z-\zeta'| < r'$ the series

$$\sum_{j=j'}^{\infty} \{v_j(z) - v_j(\zeta_0)\}$$

is absolutely and uniformly convergent, provided that the series

$$\sum_{j=j'}^{\infty} |v_j(\zeta') - v_j(\zeta_0)|$$

is convergent.

Now consider a point $\zeta$ of the domain $D$ such that $\zeta \neq \zeta_0$. Join $\zeta_0$ and $\zeta$ by a polygonal line $L$ lying in $D$ and let $P$ be a positive number such that, for each point $a \in L$, the circle $\bar{C}_a: |z-a| \leq P$ belongs to $D$. Choose a system of points $a_i$ $(i = 0,1,2,\ldots,m)$ on $L$, in the order from $\zeta_0$

to $\zeta$, such that

$$|a_i - a_{i-1}| < \rho \quad (i = 1,2,\ldots,m)$$

$$a_0 = \zeta_0, \; a_m = \zeta.$$

By a result obtained above, the series (24) is absolutely convergent in the circle $C_{a_0} = C_{\zeta_0} : |z-\zeta_0| < \rho$, which contains the point $a_1$. Next consider the circle $C_{a_1} : |z-a_1| < P$. Since the series (24) is absolutely convergent $\rho$ at $z = a_1$, it follows from a preceeding result, that the series (24) is absolutely and uniformly convergent in the circle $C_{a_1}$ which contains the point $a_2$. So the series (24) is also absolutely and uniformly convergent in the circle $C_{a_2} : |z-a_2| < \rho$. In this way, we see finally that the series (24) is absolutely and uniformly convergent in the circle $C_{a_m} = C_\zeta : |z-\zeta| <$ .

Thus the series (24) is locally uniformly convergent in the domain D and therefore defines a function $v(z)$ holomorphic in D. Moreover in D we have

$$v'(z) = \sum_{j=0}^{\infty} \{v_j(z) - v_j(\zeta_0)\}' = U_0(z) + \sum_{j=1}^{\infty} \{U_j(z) - P_j(z)\} = u(z).$$

It remains to prove that the points $z_j$ $(j = 0,1,2,\ldots)$ are poles of the function $v(z)$. To see this, consider a point $z_{j_0}$ and a domain $0 < |z-z_{j_0}| < r$ containing none of the points $z_j$ $(j = 0,1,2,\ldots)$. Take a point $\zeta'$ such that $|\zeta'-z_{j_0}| = \frac{r}{4}$. Then the circle $|z-\zeta'| \leq \frac{r}{2}$ contains none of the points $z_j$ $(j \neq j_0)$. Moreover the series (24) being absolutely convergent at $z = \zeta'$, the series

$$\sum_{j=j_0+1}^{\infty} |v_j(\zeta') - v_j(\zeta_0)|$$

is convergent. Consequently by a preceeding result, the series

$$\phi(z) = \sum_{j=j_0+1}^{\infty} \{v_j(z) - v_j(\zeta_0)\}$$

is absolutely and uniformly convergent in the circle $\gamma' : |z-\zeta'| < \frac{r}{2}$ and defines a holomorphic function $\phi(z)$ in $\gamma'$. Since $\gamma'$ contains the circle $c : |z-z_{j_0}| < \frac{r}{4}$, $\phi(z)$ is holomorphic in $c$. On the other hand, $c$ contains none of the points $z_j$ $(j = 0,1,\ldots,j_0-1)$, hence the function

$$\psi(z) = \sum_{j=0}^{j_0-1} \{v_j(z) - v_j(\zeta_0)\}$$

is holomorphic in $c$. Hence we have

$$v(z) = v_{j_0}(z) - v_{j_0}(\zeta_0) + \psi(z) + \phi(z)$$

in the domain $0 < |z-z_{j_0}| < \frac{r}{4}$. This shows clearly that $z_{j_0}$ is a pole of the function $v(z)$.

In the above, we have proved that, for a fixed $k$, a meromorphic function $C_k$ can be found satisfying the identity (14), if the function $g_k$ has an infinite number of poles. The proof is much simpler, when the number of poles of the function $g_k$ is finite.

In what follows, we apply the formula (37) to the study of the comparative growth of the functions $f(z)$ and $F(z) = L(f)$ defined by (1).

THEOREM 1. Let $f(z)$ and $\psi_k(z)$ $(k = 1,2,\ldots,P, P \geq 1)$ be meromorphic functions such that $f(z)$ is not identically equal to zero and $\psi_k(z)$ $(k = 1,2,\ldots,p)$ are linearly independent. Let $F(z) = L(f)$ be the meromorphic function defined by (1). Then for $r > 1$, we have the inequality

$$T(r,F) \leq T(r,f) + P\overline{N}(r,f) + A\sum_{k=1}^{p}\{N(r,\psi_k) + N(r,\frac{1}{\psi_k})\} + S(r), \tag{25}$$

where $S(r)$ satisfies for $1 < r < R$, the inequality

$$S(r) < a\{\log^+ T(R,f) + \sum_{k=1}^{p} \log^+ T(R,\psi_k)\} + b\log^+\frac{1}{R-r} + c\log R + d, \tag{26}$$

$A$, $a$, $b$, $c$, $d$ being positive constants.

Proof: We have

$$T(r,F) = m(r,F) + N(r,F) = m(r,f\,\frac{F}{f}) + N(r,F) \leq m(r,f) + m(r,\frac{F}{f}) + N(r,F).$$

By the inequality (17) in the paper [2], we have

$$m(r,\frac{F}{f}) \leq \sum_{k=1}^{p} m(r,\frac{f^{(k)}}{f}) + pm(r,D_0) + \sum_{k=1}^{p} m(r,D_k)$$

$$+ p\{\sum_{k=1}^{p} N(r,\frac{1}{\psi_k}) - \sum_{k=1}^{p} N(r,\psi_k) + N(r,A_0) - N(r,\frac{1}{A_0})\} + h,$$

where $h$ is a constant. On the other hand, by the inequality (13) in the paper [2], we have

$$N(r,F) \le N(r,f) + p\overline{N}(r,f) + \sum_{k=1}^{p} \{N(r,\psi_k) + p\overline{N}(r,\psi_k)\}$$

$$+ N(r,\frac{1}{A_0}) + \lambda \log r. \tag{27}$$

where $\lambda$ is a constant.

Hence we have for $r > 1$,

$$T(r,F) \le T(r,f) + p\overline{N}(r,f) + \phi(r) + S(r), \tag{28}$$

where

$$\phi(r) = p\{\sum_{k=1}^{p} (\overline{N}(r,\psi_k) + N(r,\frac{1}{\psi_k})) + N(r,A_0)\}, \tag{29}$$

$$S(r) = \sum_{k=1}^{p} m(r,\frac{f^{(k)}}{f}) + pm(r,D_0) + \sum_{k=1}^{p} m(r,D_k) + \lambda \log r + h. \tag{30}$$

Consider the term $N(r,A_0)$. Each term of the determinant $A_0$ is of the form

$$\Pi = \pm\psi_{i_1}\psi'_{i_2}\psi''_{i_3} \cdots \psi_{i_p}^{(p-1)},$$

where $(i_1,i_2,i_3,\ldots,i_p)$ is a permutation of $(1,2,3,\ldots,p)$. We have

$$N(r,\Pi) \le \sum_{k=1}^{p} N(r,\psi_{i_k}^{(k-1)}) = \sum_{k=1}^{p} \{N(r,\psi_{i_k}) + (k-1)\overline{N}(r,\psi_{i_k})\}$$

$$\le \sum_{k=1}^{p} N(r,\psi_k) + (p-1) \sum_{k=1}^{p} \overline{N}(r,\psi_k),$$

and hence

$$N(r,A_0) \le p!\{\sum_{k=1}^{p} N(r,\psi_k) + (p-1) \sum_{k=1}^{p} \overline{N}(r,\psi_k)\}. \tag{31}$$

From (29) and (31), it follows that

$$\phi(r) \le A \sum_{k=1}^{p} \{N(r,\psi_k) + N(r,\frac{1}{\psi_k})\}, \tag{32}$$

where $A$ is a positive constant. On the other hand, $D_k (0 \le k \le p)$ being the determinant obtained by omitting the $(p - k + 1)$-th row of the matrix

$$\begin{pmatrix} 1 & 1 & \cdots & 1 \\ \frac{\psi_1'}{\psi_1} & \frac{\psi_2'}{\psi_2} & \cdots & \frac{\psi_p'}{\psi_p} \\ \frac{\psi_1''}{\psi_1} & \frac{\psi_2''}{\psi_2} & \cdots & \frac{\psi_p''}{\psi_p} \\ \cdots & \cdots & \cdots & \cdots \\ \frac{\psi_1^{(p)}}{\psi_1} & \frac{\psi_2^{(p)}}{\psi_2} & & \frac{\psi_p^{(p)}}{\psi_p} \end{pmatrix},$$

we have

$$m(r,D_k) \le p!\{\sum_{k=1}^{p} m(r,\frac{\psi_k'}{\psi_k}) + m(r,\frac{\psi_k''}{\psi_k}) +\dots+ m(r,\frac{\psi_k^{(p)}}{\psi_k})\} + \log p!$$

$$(k = 0,1,2,\dots,p). \tag{33}$$

Now from a well known theorem of Nevanlinna on $m(r,\frac{g'}{g})$ [1], it is easily deduced that if $g(z)$ is a meromorphic function non identically equal to zero and if $n$ is a positive integer, then there are positive constants $a_j$ $(j=1,2,3,4)$ such that for $1 < r < R$, we have

$$m(r,\frac{g^{(n)}}{g}) < \alpha_1 \overset{+}{\log} T(R,g) + \alpha_2 \overset{+}{\log} \frac{1}{R-r} + \alpha_3 \log R + \alpha_4.$$

Then from (30) and (33), it is clear that $S(r)$ satisfies for $1 < r < R$ an inequality of the form (26). Thus Theorem 1 is proved.

THEOREM 2. Let $f(z)$ be a meromorphic function and $\psi_k(z)$ $(k = 1,2,\dots,p;\ p \ge 1)$ by $p$ linearly independent meromorphic functions. Let $F(z) = L(f)$ be the meromorphic function defined by (1). Then we can find a number $r_1$ $(r_1 > 1)$ such that for $r > r_1$, $\lambda > 1$, we have the inequality

$$T(r,f) < Bp\,\frac{\lambda+1}{\lambda-1} \log \frac{e(\lambda+1)}{\lambda-1} \{T(\lambda r,F) + b_1 \sum_{k=1}^{p} T(\lambda r,\psi_k) + b_2 \overset{+}{\log} \frac{2}{(\lambda-1)r}$$

$$+ b_3 \log(\lambda r) + b_4\}, \tag{34}$$

where $B$ is a positive numerical constant and $b_j$ $(j = 1,2,3,4)$ are positive constants (independent of $\lambda$).

Proof: By the formula (3), we have for $r > 1$,

$$T(r,f) \le \sum_{k=1}^{p} T(r,C_k) + \sum_{k=1}^{p} T(r,\psi_k) + \log p \tag{35}$$

$k$ being fixed, distinguish two cases: $1^0$ $C_k(z)$ is a rational function. Then

$$T(r,C_k) = O(\log r). \tag{36}$$

$2^0$ $C_k(z)$ is not a rational function. Then by the theorem 1 in the paper [4], there exists a positive number $r_0$ $(r_0 > 1)$ such that for $\lambda > 1$, $r > r_0$, we have

$$T(r,C_k) < B\frac{\lambda}{\lambda-1}\log\frac{e\lambda}{\lambda-1}T(\lambda r,C_k'). \tag{37}$$

From (4) we have

$$T(r,C_k') \leq T(r,\Phi_k) + T(r,\frac{1}{A_0}) + T(r,F)$$

$$= T(r,\Phi_k) + T(r,A_0) + T(r,F) + \ell, \tag{38}$$

where $\ell$ is a constant. We have

$$T(r,A_0) = m(r,A_0) + N(r,A_0).$$

From (31),

$$N(r,A_0) \leq p(p!)\sum_{k=1}^{p} N(r,\psi_k).$$

On the other hand, from the identity

$$A_0 = (\psi_1\psi_2\cdots\psi_p)D_0,$$

$$m(r,A_0) \leq \sum_{k=1}^{p} m(r,\psi_k) + m(r,D_0).$$

Hence

$$T(r,A_0) \leq p(p!)\sum_{k=1}^{p} T(r,\psi_k) + m(r,D_0).$$

Then making use of (33), we see that for $1 < r < R$ we have the inequality

$$T(r,A_0) \leq p(p!)\sum_{k=1}^{p} T(r,\psi_k) + a_1\sum_{k=1}^{p}\log^+T(R,\psi_k) + a_2\log^+\frac{1}{R-r}$$

$$+ a_3\log R + a_4, \tag{39}$$

where $a_j$ $(j = 1,2,3,4)$ are positive constants. Since the determinaint $\Phi_k$ has the same form as $A_0$, we see that the inequality (39) still holds, when in it, $A_0$ is replaced by $\Phi_k$.

Now consider two numbers $\lambda > 1$ and $r > r_0$ and set

$$\mu = \frac{\lambda+1}{2} .$$

By (37) we have

$$T(r,C_k) < B \frac{\mu}{\mu-1} \log \frac{e\mu}{\mu-1} T(\mu r,C'_k),$$

and by (38) and (39), we have

$$T(\mu r,C'_k) \leq T(\mu r,F) + \ell + 2\{p(p!) \sum_{k=1}^{p} T(\mu r,\psi_k)$$

$$+ a_1 \sum_{k=1}^{p} \log^+ T(\lambda r,\psi_k) + a_2 \log^+ \frac{1}{\lambda r-\mu r} + a_3 \log(\lambda_r) + a_4\}.$$

Now it is easy to see that in either of the two cases considered above, we can find a number $r'_0$ $(r'_0 > 1)$ such that for $r > r'_0$, $\lambda > 1$, we have

$$T(r,C_k) < B \frac{\lambda+1}{\lambda-1} \log \frac{e(\lambda+1)}{\lambda-1} \{T(\lambda r,F) + a'_1 \sum_{k=1}^{p} R(\lambda r,\psi_k)$$

$$+ a'_2 \log^+ \frac{2}{(\lambda-1)r} + a'_3 \log (\lambda_r) + a'_4\} + K \log r, \tag{40}$$

where $K$ is a positive constant.

Finally from (35) and (40), we conclude that we can find a number $r_1$ having the required property in theorem 2.

We are going to deduce some consequences from the theorems 1 and 2. First we give a definition.

DEFINITION 1. Let $f(z)$ be a transcendental meromorphic function of finite order $\rho$ and $U(r)$ $(r \geq r_0)$ a positive continuous function tending to $+\infty$ with $r$ and having the following properties:

$1^0$ $\displaystyle\lim_{r\to+\infty} \frac{\log U(r)}{\log r} = \rho.$

$2^0$ There exists a number $\lambda > 1$ such that $U(\lambda r) = O\{U(r)\}$.

$3^0$ $\displaystyle 0 < \overline{\lim_{r\to+\infty}} \frac{T(r,f)}{U(r)} < +\infty.$

It is known that such a function $U(r)$ exists [5] [6] [7]. We denote by $S(f,U)$ the set of meromorphic functions $\psi(z)$ satisfying the condition

$$T(r,\psi) = o\{U(r)\}. \tag{41}$$

The set $S(f,U)$ contains in particular the meromorphic functions $\psi(z)$

satisfying the condition

$$T(r,\psi) = o\{T(r,f)\}$$

and, in the case $\rho > 0$, the meromorphic functions $\psi(z)$ is of order less than $\rho$.

THEOREM 3. Let $f(z)$ be a transcendental meromorphic function of finite order $\rho$ and $U(r)$ $(r \geq r_o)$ a positive continuous function tending to $+\infty$ with $r$ and having the properties $1^0$, $2^0$, $3^0$ in definition 1. Let $\psi_k(z)$ $(k = 1,2,\ldots,p;\ p \geq 1)$ be $p$ linearly independent meromorphic functions of the set $S(f,U)$. Then the meromorphic function $F(z) = L(f)$ defined by (1) is transcendental and has the same order $\rho$.

Proof: Let us first prove that $F(z)$ is a transcendental meromorphic function of order not less than $\rho$. In fact, by theorem 2, for $r > r_1$, we have the inequality (34), where $\lambda$ is taken to be the number occuring in the property $2^0$ in definition 1 and kept fixed. Now suppose that $F(z)$ is a rational function. Then

$$T(\lambda r,F) = O(\log r).$$

Since $\psi_k(z) \in S(f,U)$ $(k = 1,2,\ldots,p)$, the equality

$$\frac{T(\lambda r,\psi_k)}{U(r)} = \frac{T(\lambda r,\psi_k)}{U(\lambda r)}\,\frac{U(\lambda r)}{U(r)}$$

and the property $2^0$ in definition 1 show that

$$T(\lambda r,\psi_k) = o\{U(r)\} \quad (k = 1,2,\ldots,p). \tag{42}$$

On the other hand, since $f(z)$ is transcendental, the property $3^0$ in definition 1 implies

$$\log r = o\{U(r)\}. \tag{43}$$

It follows then from (34) that

$$T(r,f) = o\{U(r)\} \tag{44}$$

which is incompatible with the property $3^0$ in definition 1. So $F(z)$ is a transcendental meromorphic function. In the same way we see that the order of $F(z)$ is not less than $\rho$, in case $\rho > 0$. For if the order of $F(z)$ is less than $\rho$, then by the property $1^0$ in the definition 1, we have

$$T(r,F) = o\{U(r)\}$$

and get again (44).

Next we apply theorem 1 to show that the order of $F(z)$ is not greater than $\rho$. In the equality (26) taking $R = \lambda r$, where $\lambda$ is the number occuring in the property $2^0$ in definition 1, and then making use of the properties $2^0$ and $3^0$ in definition 1, we deduce easily from (25) and (26), the inequalities

$$T(r,F) < (p+1)T(r,f) + o\{U(r)\} < KU(r),$$

where $K$ is a positive constant, provided that $r$ is sufficiently large. Then by the property $1^0$ in definition 1, evidently the order of $F(z)$ is not greater than $\rho$.

DEFINITION 2. Let $f(z)$ be a transcendental meromorphic function. We denote by $\sigma(r)$ the set of meromorphic functions $\psi(z)$ satisfying the following condition: There exist a number $\lambda > 1$ ($\lambda$ depends on $\psi(z)$) such that

$$T(\lambda r,\psi) = o\{T(r,f)\}. \tag{45}$$

The set $\sigma(f)$ contains in particular the rational functions. If there is a number $\mu > 1$ such that $T(\mu r,f) = o\{T(r,f)\}$, then $\sigma(f)$ contains the meromorphic functions $\psi(z)$ such that $T(r,\psi) = o\{T(r,f)\}$. Besides in the case that $f(z)$ is of positive order and of normal growth, namely

$$\lim_{r\to+\infty} \frac{\log T(r,f)}{\log r} = \rho, \ 0 < \rho \leq +\infty,$$

then $\sigma(f)$ contains the meromorphic functions $\psi(z)$ of order less than $\rho$.

THEOREM 4. Let $f(z)$ be a transcendental meromorphic function and $\psi_k(z)$ $(k = 1,2,\ldots,p;\ p \geq 1)$ be $p$ linearly independent meromorphic functions of the set $\sigma(f)$. Then the meromorphic function $F(z) = L(f)$ defined by (1) is transcendental and the functions $f(z)$ and $F(z)$ have the same order and the same lower order.

Proof: By hypothesis, there are numbers $\lambda_k > 1$ $(k = 1,2,\ldots,p)$ such that

$$T(\lambda_k r,\psi_k) = o\{T(r,f)\} \ (k = 1,2,\ldots,p).$$

In the inequality (34), taking $\lambda = \min(\lambda_1,\lambda_2,\ldots,\lambda_p)$, we see that for $r > r_1$,

$$T(r,f) < A\{T(\lambda r,F) + o(T(r,f))\},$$

where $A$ is a positive constant. Consequently, when $r$ is sufficiently large, we have

$$T(r,f) < 2AT(\lambda r,F). \tag{46}$$

Denote by $\rho_f$ and $\rho_F$ respectively the order of $f(z)$ and that of $F(z)$, and denote by $\rho'_f$ and $\rho'_F$ respectively the lower order of $f(z)$ and that of

$F(z)$. Evidently (46) implies that $F(z)$ is transcendental and

$$\rho_f \leq \rho_F, \ \rho'_f \leq \rho'_F. \tag{47}$$

Next we apply theorem 1 to obtain the inequalities

$$\rho_F \leq \rho_f, \ \rho'_F \leq \rho'_f. \tag{48}$$

In fact, in the inequality (26), taking $R = \lambda r$, then we see from (25) and (26) that, when $r$ is sufficiently large, we have

$$T(r,F) < (p+2)T(\lambda r,f),$$

which implies (48). Therefore $\rho_f = \rho_F$ and $\rho'_f = \rho'_F$.

DEFINITION 3. Let $f(z)$ be a transcendental meromorphic function. We denote by $\hat{\sigma}(f)$ the union of the set $\sigma(f)$ and the value $\infty$:

$$\hat{\sigma}(f) = \sigma(f) \cup (\infty).$$

DEFINITION 4. Let $f(z)$ be a transcendental meromorphic function. To each element $\phi \in \hat{\sigma}(f)$ we associate a number $\delta(\phi,f)$ defined as follows:

$$\delta(\infty,f) = \underline{\lim}_{r\to+\infty} \frac{m(r,f)}{T(r,f)} = 1 - \overline{\lim}_{r\to+\infty} \frac{N(r,f)}{T(r,f)}$$

and, $\phi(z)$ being a meromorphic function of the set $\sigma(f)$,

$$\delta(\phi,f) = \underline{\lim}_{r\to+\infty} \frac{m(r,\frac{1}{f-\phi})}{T(r,f)} = 1 - \overline{\lim}_{r\to+\infty} \frac{N(r,\frac{1}{f-\phi})}{T(r,f)} .$$

We have always

$$0 \leq \delta(\phi,f) \leq 1.$$

THEOREM 5. Let $f(z)$ be a transcendental meromorphic function of finite order. Suppose that there exists an element $\phi_0 \in \hat{\sigma}(f)$ satisfying the following condition: There is a number $\lambda_0 > 1$ such that

$$N(\lambda_0 r, \frac{1}{f-\phi_0}) = o\{T(r,f)\}. \tag{49}$$

Then for any finite number of distinct elements $\phi_j (j = 1,2,\ldots,q;\ q \geq 2)$ of the set $\hat{\sigma}(f)$, we have

$$\sum_{j=1}^{q} \delta(\phi_j,f) \leq 2 - \underline{\lim}_{r\to+\infty} \frac{N(r,\Lambda) + N(r,\frac{1}{\Lambda})}{T(r,\Lambda)} , \tag{50}$$

where $\Lambda = \Lambda(z)$ is a transcendental meromorphic function such that $f(z)$ and $\Lambda(z)$ have the same order and the same lower order.

In this theorem, if $\phi_0 = \infty$, we define

$$N(\lambda_0 r, \frac{1}{f-\phi_0}) = N(\lambda_0 r, f);$$

two elements $\phi$ and $\psi$ of the set $\hat{\sigma}(f)$ are said to be distinct, if $\phi \not\equiv \psi$.

Theorem 5 is proved in several steps.

1) $\phi_0 = \infty$. In the case, the condition (49) is

$$N(\lambda_0 r, f) = o\{T(r,f)\}. \tag{51}$$

Let $\phi_j(z)$ $(j = 1,2,\ldots,q;\ q \geq 2)$ be distinct elements of the set $\sigma(f)$. Then we can find a positive integer $p$ $(1 \leq p \leq q)$ such that among the $q$ meromorphic functions $\phi_j(z)$ $(j = 1,2,\ldots,q)$, there are $p$ of them $\psi_k(z) = \phi_{j_k}(z)$ $(k = 1,2,\ldots,p)$ satisfying the following conditions:

$1^0$ The functions $\psi_k(z)$ $(k = 1,2,\ldots,p)$ are linearly independent.

$2^0$ Each function $\phi_j(z)$ $(1 \leq j \leq q)$ is a linear combination of $\psi_k(z)$ $(k = 1,2,\ldots,p)$ with constant coefficients, namely

$$\phi_j(z) = \sum_{k=1}^{p} C_{jk}\psi_k(z).$$

Let $L(f)$ be defined by (1). By some results obtained in the paper [2], we have the inequality

$$\begin{aligned}\sum_{j=1}^{q} m(r,\frac{1}{f-\phi_j}) &\leq m\{r,\frac{1}{L(f)}\} + \sum_{j=1}^{q} m\{r,\frac{L(f-\phi_j)}{f-\phi_j}\} \\ &+ q \sum_{1\leq j_1 \leq j_2 \leq q} m(r,\frac{1}{\phi_{j_1}-\phi_{j_2}}) + h,\end{aligned} \tag{52}$$

where $h$ is a constant.

By the inequality (18) in the paper [2], we have

$$\begin{aligned}m\{r,\frac{L(f-\phi_j)}{f-\phi_j}\} &\leq \sum_{k=1}^{p} m\{r,\frac{(f-\phi_j)^{(k)}}{f-\phi_j}\} + pm(r,D_0) + \sum_{k=1}^{p} m(r,D_k) \\ &+ p\sum_{k=1}^{p} N(r,\frac{1}{\psi_k}) + N(r,A_0)\} + h',\end{aligned}$$

where $h'$ is a constant. Since $f(z)$ is of finite order, we see, by making use of (33), that

$$\sum_{k=1}^{p} m\{r,\frac{(f-\phi_j)^{(k)}}{f-\phi_j}\} + pm(r,D_0) + \sum_{k=1}^{p} m(r,D_k) = O(\log r).$$

On the other hand, since $\phi_j(z) \in \sigma(f)$ $(j = 1,2,\ldots,q)$, there are numbers $\lambda_j > 1$ $(j = 1,2,\ldots,q)$ such that

$$T(\lambda_j r,\phi_j) = o\{T(r,f)\} \ (j = 1,2,\ldots,q).$$

Set

$$\lambda = \min(\lambda_0,\lambda_1,\lambda_2,\ldots,\lambda_q),$$

where $\lambda_0 > 1$ is the number occuring in (51). Then, making use of (31), we see that

$$N(r,\frac{1}{\psi_k}) = o\{T(\frac{r}{\lambda},f)\},\ N(r,A_0) = o\{T(\frac{r}{\lambda},f)\},\ m(r,\frac{1}{\phi_{j_1}-\phi_{j_2}}) = o\{T(\frac{r}{\lambda},f)\}.$$

Consequently from (52), we have

$$\sum_{j=1}^{p} m(r,\frac{1}{f-\phi_j}) \leq m(r,\frac{1}{F}) + o\{T(\frac{r}{\lambda},f)\}, \tag{53}$$

where $F = F(z) = L(f)$.

Applying theorem 1 with $R = 2r$ and by (51), we get

$$T(r,f) \leq T(r,f)\{1 + o(1)\}. \tag{54}$$

On the other hand, applying theorem 2, we see that, when $r$ is sufficiently large, we have

$$T(r,f) < 2Bp\,\frac{\lambda+1}{\lambda-1}\log\frac{e(\lambda+1)}{\lambda-1}\,T(\lambda r,F). \tag{55}$$

we also need the inequality

$$N(r,F) \leq o\{T(\frac{r}{\lambda},f)\} \tag{56}$$

deduced from (27) and (39).

Now multiplying both sides of the inequality (53) by $\frac{1}{T(r,F)}$ and using (54) and (55), we have

$$\sum_{j=1}^{q} \varliminf_{r\to+\infty} \frac{m(r,\frac{1}{f-\phi_j})}{T(r,f)} \leq \varliminf_{r\to+\infty} \frac{1}{T(r,f)} \sum_{j=1}^{q} m(r,\frac{1}{f-\phi_j})$$

$$\leq \varliminf_{r\to+\infty} \frac{1}{T(r,F)} \sum_{j=1}^{q} m(r,\frac{1}{f-\phi_j}) \leq \varliminf_{r\to+\infty} \frac{m(r,\frac{1}{F})}{T(r,F)}.$$

Hence by definition 4,

$$\sum_{j=1}^{q} \delta(\phi_j,f) \leq 1 - \varlimsup_{r\to+\infty} \frac{N(r,\frac{1}{F})}{T(r,F)}.$$

This inequality can also be written in the form

$$\sum_{j=1}^{q} \delta(\phi_j,f) \leq 1 - \varlimsup_{r\to+\infty} \frac{N(r,F)+N(r,\frac{1}{F})}{T(r,F)}, \tag{57}$$

because from (55) and (56) we have

$$\lim_{r\to+\infty} \frac{N(r,F)}{T(r,F)} = 0.$$

Since $\delta(\infty,F) = 1$, by (51), the inequaltiy (57) is the same as

$$\delta(\infty,f) + \sum_{j=1}^{q} \delta(\phi_j,f) \leq 2 - \overline{\lim_{r\to+\infty}} \frac{N(r,F)+N(r,\frac{1}{F})}{T(r,F)}. \tag{58}$$

By theorem 4, $F(z)$ is a transcendental meromorphic function and the functions $f(z)$ and $F(z)$ have the same order and the same lower order, theorem 5 is thus proved in the case $\phi_0 = \infty$, with $\Lambda(z) = F(z)$.

2) $\phi_0 = 0$. In this case, the condition (49) is

$$N(\lambda_0 r,\frac{1}{f}) = o\{T(r,f)\}.$$

Consider the function

$$g(z) = \frac{1}{f(z)}.$$

We have

$$T(r,f) = T(r,g) + \alpha,$$

where $\alpha$ is a constant. Evidently $g(z)$ is also a transcendental meromorphic function and $f(z)$ and $g(z)$ have the same order and the same lower order. Moreover we have

$$N(\lambda_0 r,g) = o\{T(r,g)\}.$$

It is easy to see that

$$\delta(\infty,g) = \delta(0,f),\ \delta(0,g) = \delta(\infty,f) \tag{59}$$

and if $\phi \in \sigma(f)$, $\phi \not\equiv 0$, then $\frac{1}{\phi} \in \sigma(g)$ and by the identity

$$\frac{1}{f-\phi} = -\frac{1}{\phi}\{1 + \frac{1}{\phi}\frac{1}{g-\frac{1}{\phi}}\},$$

we see that

$$\delta(\frac{1}{\phi},g) = \delta(\phi,f) \tag{60}$$

Now let $\phi_j (j = 1,2,\ldots,q;\ q \geq 2)$ be any finite number of distinct elements of the set $\hat{\sigma}(f)$. Then $\frac{1}{\phi_j}$ $(j = 1,2,\ldots,q)$ are distinct elements of $\hat{\sigma}(g)$,

and by (59) and (60), we have

$$\sum_{j=1}^{q} \delta(\phi_j,f) = \sum_{j=1}^{q} \delta(\frac{1}{\phi_j},g) \leq 2 - \overline{\lim_{r\to+\infty}} \frac{N(r,G)+N(r,\frac{1}{G})}{T(r,G)}$$

where $G(z)$ is a transcendental meromorphic function whose order and lower order are respectively equal to the order and lower order of $f(z)$. So theorem 5 is also true in the case $\phi_0 = 0$.

3) $\phi_0$ is distinct from $\infty$ and $0$. In this case, we consider the function

$$h(z) = f(z) - \phi_0(z).$$

We have

$$T(r,f) = T(r,h) + o\{T(r,f)\}.$$

It follows that $h(z)$ is also a transcendental meromorphic function whose order and lower order are respectively equal to the order and lower order of $f(z)$. We have

$$N(\lambda_0 r, \frac{1}{h}) = o\{T(r,h)\}.$$

It is easy to see that

$$\delta(\infty,h) = \delta(\infty,f), \ \delta(0,h) = \delta(\phi_0,f) \tag{61}$$

and that if $\phi \in \sigma(f)$, $\phi \not\equiv \phi_0$, then $\phi - \phi_0 \in \sigma(h)$ and

$$\delta(\phi - \phi_0, h) = \delta(\phi, f). \tag{62}$$

From (61) and (62), we conclude that theorem 5 is also true in the case $\phi_0 \not\equiv \infty, 0$.

Theorem 5 is now completely proved.

By a classical theorem of Nevanlinna [1], if $f(z)$ is a transcendental meromorphic function of finite non-integral order $\rho$, then

$$\varliminf_{r\to+\infty} \frac{N(r,f)+N(r,\frac{1}{f})}{T(r,f)} \geq k(\rho), \tag{63}$$

where $k(\rho) > 0$ is a number depending only on $\rho$.

Then from theorem 5 we deduce the following consequence:

COROLLARY 1. Let $f(z)$ be a transcendental meromorphic function of finite non-integral order $\rho$. Suppose that there exists an element $\phi_0 \in \hat{\sigma}(f)$ satisfying the condition (49). Then for any finite number of distinct elements $\phi_j (j = 1,2,\ldots,q;\ q \geq 2)$ of the set $\hat{\sigma}(f)$, we have

$$\sum_{j=1}^{q} \delta(\phi_j, f) \leq 2 - k(\rho).$$

In particular for entire functions, we have the following corollary:

COROLLARY 2. Let $f(z)$ be a transcendental entire function of finite non-integral order $\rho$. Then for any finite number of distinct rational functions $R_j(z)(j = 1,2,\ldots,q;\ q \geq 2)$, we have

$$\sum_{j=1}^{q} \delta(R_j,f) \leq 1 - k(\rho).$$

Concerning the inequality (63) see also the works [8], [9], [10], and [11].

REFERENCES

[1] Nevanlinna, R., Le theorem de Picard-Borel et la théorie des fonctions méromorphes, Paris (1929).

[2] Chuang, Chi-tai, Une généralisation d'une inégalité de Nevanlinna, Scientia Sinica, 13(1964), 887-895.

[3] Goursat, E., Cours d'analyse mathématique, Tome II, Paris (1933).

[4] Chuang, Chi-tai, Sur la comparaison de la croissance d'une fonction méromorphe et de celle de sa dérivée, Bull. Sci. Math., 75(1951), 1-20.

[5] ______, Sur les fonctions-types, Scientia Sinica, 10(1961).

[6] ______, Singular directions of meromorphic functions (in Chinese), Beijing (1982).

[7] Valiron, G., Directions de Borel des fonctions méromorphes, Mém. Sci. Math., Fasc. 89 (1938).

[8] Hayman, W.K., Meromorphic functions, Oxford (1964).

[9] Edrei, A. and Fuchs, W.H.J., On the growth of meromorphic functions with several deficient values, Trans. Amer. Math. Soc., 93(1959), 292-328.

[10] ______, The deficiences of meromorphic functions of order less than one, Duke Math. J., 27(1960), 233-249.

[11] Edrei, A., The deficiences of meromorphic functions of finite lower order, Duke Math. J., 31(1964), 1-21.

DEPARTMENT OF MATHEMATICS
PEKING UNIVERSITY
Beijing

Contemporary Mathematics
Volume 48, 1985

# DISTRIBUTIONS OF THE VALUES OF MEROMORPHIC FUNCTIONS*

Chi-tai Chuang and Lo Yang

Since the fourth decade of this century several Chinese mathematicians, under the influence of Hiong, began to do research work on meromorphic functions. In recent years, Chinese mathematicians have made various contributions to the value distribution theory of meromorphic functions. The present article is a survey of these contributions which may be classified as follows:

1. Generalization of the second fundamental theorem of Nevanlinna and deficient functions.
2. Number of deficient values.
3. Deficient values and Borel directions.
4. Distribution of Borel directions.
5. Common Borel directions of a meromorphic function and its derivatives.
6. Extensions of Miranda theorem on normal families.
7. Asymptotic values and asymptotic paths.
8. Meromorphic functions in the unit disk or in an angle.
9. Algebroid functions.
10. Generalizations of Malmquist theorem.
11. Growth of functions.

## 1. GENERALIZATION OF THE SECOND FUNDAMENTAL THEOREM OF NEVANLINNA AND DEFICIENT FUNCTIONS

In the theory of meromorphic functions the importance of the second fundamental theorem of Nevanlinna:

$$(q-2)T(r,f) < \sum_{j=1}^{q} N(r,a_j) - N_1(r) + S(r) \tag{1.1}$$

is well known. In 1929, Nevanlinna [1] proposed to generalize this theorem in replacing the values $a_j$ $(j = 1,2,\ldots,q)$ by meromorphic functions $\phi_j(z)$ $(j=1,2,\ldots,q)$ satisfying the condition:

$$T(r,\phi_j) = o\{T(r,f)\} \quad (j=1,2,\ldots,q). \tag{1.2}$$

*The present article was originally planned to be written by Chi-tai Chuang, Lo Yang, and Kuan-heo Chang. Unfortunately, because of sickness, Kuan-heo Chang had been unable to participate in this joint work.


0271-4132/85 $1.00 + $.25 per page

Nevanlinna himself solved this problem in the case $q = 3$. In the general case $q \geq 3$, this problem is easier to deal with when the $\phi_j (j = 1,2,\ldots,q)$ are polynomials, and this was done by Dufresnoy [2]. However when the $\phi_j (j = 1,2,\ldots,q)$ are subjected only to the condition (1.2), this problem is more difficult. In order to treat it, we naturally follow the procedure used by Nevanlinna in his proof of the inequality (1.1) by introducing the auxiliary function

$$F(z) = \sum_{j=1}^{q} \frac{1}{f(z)-\phi_j(z)}$$

and finding a lower bound and an upper bound of $m(r,F)$. To get a lower bound of $m(r,F)$ presents no difficulty, but it is rather hard to get an upper bound of $m(r,F)$. Chuang [3] overcame this difficulty by considering $p$ linearly independent meromorphic functions $\psi_k(z) (k = 1,2,\ldots,p)$ satisfying the condition

$$T(r,\psi_k) = o\{T(r,f)\} \quad (k=1,2,\ldots,p)$$

and introducing the linear operator

$$L(f) = (-1)^p \frac{\Delta(f,\psi_1,\psi_2,\ldots,\psi_p)}{\Delta(\psi_1,\psi_2,\ldots,\psi_p)},$$

where $\Delta(\psi_1,\psi_2,\ldots,\psi_p)$ and $\Delta(f,\psi_1,\psi_2,\ldots,\psi_p)$ are respectively the Wronskian of $\psi_k (k=1,2,\ldots,p)$ and that of $f$, $\psi_k (k=1,2,\ldots,p)$. In particular, for $p = 1$, $\psi_1(z) = 1$, $L(f) = f'(z)$ and for $p \geq 2$, $\psi_k(z) = z^{k-1} (k=1,2,\ldots,p)$, $L(f) = f^{(p)}(z)$.

$\phi_j(z) (j=1,2,\ldots,q)$ being $q$ distinct linear combinations of $\psi_k(z)$ $(k=1,2,\ldots,p)$ with constant coefficients, Chuang obtained an upper bound of $m(r,F)$ in writing

$$F(z) = \frac{1}{L\{f(z)\}} \sum_{j=1}^{q} \frac{L\{f(z)-\phi_j(z)\}}{f(z)-\phi_j(z)}.$$

In this way he proved some theorems [3] [4] [5] which solve almost completely the problem proposed by Nevanlinna. Here we state only one of these theorems as follows:

Let $f(z)$ be a transcendental meromorphic function and $e(f)$ the set consisting of all the meromorphic functions $\phi(z)$ satisfying the condition:

$$T(r,\phi) = o\{T(r,f)\}$$

and of the constant $\phi(z) = \infty$. For an element $\phi(z)$ of $e(f)$ and a positive integer $p$, denote by $n_p(t, \frac{1}{f-\phi})$ the number of the roots of the equation $f(z) = \phi(z)$ in the disk $|z| \leq t$, each root of order $m$ being counted $\min(m,p)$ times, and set

$$N_p(r, \frac{1}{f-\phi}) = \int_0^r \frac{n_p(t, \frac{1}{f-\phi})-n_p(0, \frac{1}{f-\phi})}{t}\,dt + n_p(0, \frac{1}{f-\phi})\log r,$$

$$\lambda_p(\phi,f) = \overline{\lim_{r\to+\infty}} \frac{pN_1(r, \frac{1}{f-\phi})-N_p(t, \frac{1}{f-\phi})}{T(r,f)},$$

$$\lambda(\phi,f) = \lim_{p\to+\infty} \lambda_p(\phi,f).$$

THEOREM 1.1. If there exists an element $\phi_0(z)$ of $e(f)$ such that $\lambda(\phi_0,f) < +\infty$, then for any $q$ distinct elements $\phi_j(z)$ $(j=1,2,\ldots,q;\ q \geq 3)$ of $e(f)$, the inequality

$$\{q-2-\lambda(\phi_0,f)\}T(r,f) < \sum_{j=1}^{q} N(r, \frac{1}{f-\phi_j}) + S(r) \tag{1.3}$$

holds for $r > 1$, where the function $S(r)$ satisfies the condition: there exists a set $\sigma$ of values of $r$ of finite exterior measure such that

$$\lim_{\substack{r\to+\infty \\ r\notin\sigma}} \frac{S(r)}{T(r,f)} = 0.$$

$\phi(z)$ being an element of $e(f)$, define[1)]

$$\delta(\phi,f) = \underline{\lim}_{r\to+\infty} \frac{m(r, \frac{1}{f-\phi})}{T(r,f)} = 1 - \overline{\lim_{r\to+\infty}} \frac{N(r, \frac{1}{f-\phi})}{T(r,f)}.$$

We have $0 \leq \delta(\phi,f) \leq 1$. If $\delta(\phi,f) > 0$, then $\phi(z)$ is called a deficient function of $f(z)$ and $\delta(\phi,f)$ the deficiency of $\phi(z)$ with respect to $f(z)$. Since the inequality (1.3) implies

$$\sum_{j=1}^{q} \delta(\phi_j,f) \leq 2 + \lambda(\phi_0,f),$$

it follows that, under the condition of Theorem 1.1, the set of the deficient functions of $f(z)$ is countable and the total sum of the corresponding deficiencies does not exceed $2 + \lambda(\phi_0,f)$. In particular if $f(z)$ is a transcendental entire function, then $\lambda(\infty,f) = 0$ and so the total sum of the deficiencies of the deficient functions $\phi(z) \not\equiv \infty$ of $f(z)$ does not exceed 1.

Yang [6] extended the total spread relation which had been conjectured by Edrei and proved by Baernstein [7] respectively, to the case of deficient functions as follows.

---

1) If $\phi(z) = \infty$, then $m(r, \frac{1}{f-\phi})$ and $N(r, \frac{1}{f-\phi})$ are understood to be $m(r,f)$ and $N(r,f)$ respectively.

If $f(z)$ is a meromorphic function of finite lower order $\mu$ and $\phi_j(z)$ $(j=1,2,\ldots,q;\ 2 \le q \le \infty)$ be its deficient functions. Then we have

$$\frac{4}{\mu} \sum_{j=1}^{q} \arcsin \sqrt{\frac{\delta(\phi_j,f)}{2}} \le 2\pi. \tag{1.4}$$

Using (1.4), the following theorem can be obtained immediately.

THEOREM 1.2. If $f(z)$ is a meromorphic function of finite lower order $\mu$, then the set of its deficient functions is countable. Moreover the total sum of the corresponding deficiencies does not exceed

$$\min\{[2\mu] + 1,\ \max(1, \tfrac{\sqrt{2}}{2}\mu\pi)\}.$$

Some results on deficient values have been transferred to deficient functions. For instance, the following theorem which extended a theorem of Edrei, was proved.

THEOREM 1.3. If $f(z)$ is meromorphic of lower order $\mu < 1$, then

(i) When $\mu = 0$, $f(z)$ has one deficient function at most.

(ii) When $0 < \mu \le \frac{1}{2}$,

$$\sum_{j=1}^{q} \delta(\phi_j,f) \begin{cases} \le 1, & \text{if } q = 1 \\ \le 1 - \cos\mu\pi, & \text{if } q \ge 2. \end{cases}$$

(iii) When $\frac{1}{2} < \mu \le 1$,

$$\sum_{j=1}^{q} \delta(\phi_j,f) \le 2 - \sin\mu\pi,$$

where the equality holds if and only if $q = 2$, $\delta(\phi_1,f) = 1$ and $\delta(\phi_2,f) = 1 - \sin\mu\pi$.

Finally we mention some results of Chen [9] on Valiron deficient functions, which generalizes a theorem of Hyllengren [10] on Valiron deficient values. Chen denoted the set of meromorphic functions by $M$ and, inspired by a paper of Rauch [11], he gave a definition of the pseudo-distance $d(\phi,\psi)$ between two elements $\phi(z), \psi(z)$ of $M$ as follows: If $\phi(z) \not\equiv \psi(z)$ and the Laurent development of $\phi(z) - \psi(z)$ in the neighborhood of the point $z = 0$ is

$$\phi(z) - \psi(z) = c_s z^s + c_{s+1} z^{s+1} + \ldots, \quad c_s \ne 0.$$

then $d(\phi,\psi) = |c_s|$. If $\phi(z) \equiv \psi(z)$, then $d(\phi,\psi) = 0$. In particular, when $\phi(z), \psi(z)$ are two constants $a, b$, we have $d(\phi,\psi) = |a-b|$. Next, following Hyllengren, Chen gave the definition of a $\mu$-set of meromorphic functions. A set $F \subset M$ is called a $\mu$-set, if there exists a positive number

$\sigma$ and a sequence $\psi_n(z) \in M(n=1,2,\ldots)$ such that

$$F \subset \bigcap_{N=1}^{\infty} \bigcup_{n=N}^{\infty} D_n,$$

where $D_n$ is the set

$$D_n = \{\phi | d(\phi,\psi_n) < \exp(-\exp(\sigma n)),\ \phi \in M\}.$$

Now let $f(z)$ be a transcendental meromorphic function. A meromorphic function $\phi(z)$ satisfying the condition $T(r,\phi) = o\{T(r,f)\}$ is called a Valiron deficient function of $f(z)$, if the number

$$\Delta(\phi,f) = \overline{\lim_{r\to+\infty}} \frac{m(r, \frac{1}{f-\phi})}{T(r,f)} = 1 - \underline{\lim_{r\to+\infty}} \frac{N(r, \frac{1}{f-\phi})}{T(r,f)}$$

is positive. Consider $p$ linearly independent meromorphic functions $\psi_k(z)$ $(k=1,2,\ldots,p)$ satisfying the condition

$$T(r,\psi_k) = o\{T(r,f)\} \quad (k=1,2,\ldots,p)$$

and denote by $M(\psi_1,\psi_2,\ldots,\psi_p)$ the set of meromorphic functions of the form $\sum_{j=1}^{p} C_j\psi_j(z)$, where $C_j\ (j=1,2,\ldots,p)$ are constants. On the basis of a theorem of Chuang [3] [4], Chen proved the following

THEOREM 1.4. If the transcendental meromorphic function $f(z)$ is of finite order, then the set

$$F_\delta = \{\phi | \Delta(\phi,f) < \delta,\ \ \phi \in M(\psi_1,\psi_2,\ldots,\psi_p)\}$$

is a $\mu$-set, where $\delta$ is a number such that $0 < \delta < 1$.

In case the function $f(z)$ is of infinite order, he also proved an analogous theorem in which $\Delta(\phi,f)$ is replaced by

$$\Delta'(\phi,f) = \overline{\lim_{\substack{r\to+\infty \\ r\notin I}}} \frac{m(r, \frac{1}{f-\phi})}{T(r,f)},$$

where $I$ is a sequence of intervals of finite total length and depending only on the function $f(z)$.

## 2. NUMBER OF DEFICIENT VALUES

Let $f(z)$ be an entire function and $\delta(a,f)$ be the deficiency of the value $a$. The famous deficient relation has the form

$$\sum_{a\neq\infty} \delta(a,f) \leq 1. \tag{2.1}$$

It is natural to raise this important question: under what conditions the equality in (2.1) holds?

For this, Pfluger [12] firstly proved the following beautiful theorem. Let $f(z)$ be an entire function of finite order $\lambda$. If

$$\sum_{a\neq\infty} \delta(a,f) = 1, \tag{2.2}$$

then $\lambda$ is a positive integer and all the deficiencies are integral multiples of $\frac{1}{\lambda}$. Consequently, the number of finite deficient values of $f(z)$ is at most equal to $\lambda$. Edrei and Fuchs [13] have notably extended the Pfluger's result. They have shown that in this case $f(z)$ also has lower order $\lambda$ and all the deficient values $a_j$ of $f(z)$ are asymptotic values, i.e. $f(z) \to a_j$ as $z \to \infty$ along a continuous path $\Gamma_j$.

For every entire function $f(z)$ of finite order, it follows from a well known fact [14] that

$$\sum_{j=-\infty}^{\infty} \sum_{a\neq 0,\infty} \delta(a,f^{(j)}) \leq 1, \tag{2.3}$$

where $f^{(j)}$ denote the derivatives of order $j$ of $f$ when $j = 1,2,\ldots$ and the primitives of order $|j|$ when $j = -1, -2, \ldots$ and $f^{(0)} \equiv f$.

We may ask whether the Pfluger's and Edrei-Fuchs' results still hold, if (2.2) is replaced by the more general condition

$$\sum_{j=-\infty}^{\infty} \sum_{a\neq 0,\infty} \delta(a,f^{(j)}) = 1. \tag{2.4}$$

THEOREM 2.1. (Yang and Zhang) [15] [16]. If $f(z)$ is an entire function of finite lower order $\mu$ with (2.4), then we have

(i) $\lambda$, the order of $f(z)$, equals $\mu$ and is a positive integer.

(ii) $\sum_{j=-\infty}^{\infty} p_j \leq \mu$, where $p_j (j=0,\pm1,\pm2,\ldots)$ denote the numbers of finite and non-zero deficient values of $f^{(j)}(z)$.

(iii) Every deficient value of $f^{(j)}(z)$ $(j=0,\pm1,\pm2,\ldots;\ f^{(0)} \equiv f)$ is also an asymptotic value of $f^{(j)}(z)$.

(iv) Every deficiency of $f^{(j)}(z)$ is a multiple of $\frac{1}{\mu}$.

There are many other results on the estimate of the number of deficient values. For instance, Edrei and Fuchs proved an interesting theorem which says: If all the zeros and poles of a meromorphic function $f(z)$ are located on $q$ $(<+\infty)$ rays, then the number of finite non-zero deficient values of $f(z)$ does not exceed $q$.

Yang and Zhang extended this result to a general case. They gave an estimate on the total sum of the numbers of finite non-zero deficient values of $f(z)$ and all its derivatives. To formulate their result, the following definition is needed.

DEFINITION. Suppose that $f(z)$ is a meromorphic function of order $\lambda(0<\lambda<\infty)$, that $\arg z = \theta_0 (0 \le \theta_0 < 2\pi)$ is a ray and that $\alpha$ is a complex value. $\arg z = \theta_0$ is named a cluster line of order $\lambda$ of $f(z)$ for $\alpha$-points, if

$$\lim_{\varepsilon\to 0} \overline{\lim_{r\to\infty}} \left\{ \frac{\log n(r,\theta_0,\varepsilon,f=\alpha)}{\log r} \right\} = \lambda.$$

When $\alpha = 0$ or $\infty$, $\arg z = \theta_0$ is named a cluster line for zeros or poles.

THEOREM 2.2 (Yang and Zhang) [17]. Let $f(z)$ be a meromorphic function of order $\lambda(0<\lambda<\infty)$. Suppose that the number of cluster lines for zeros or poles of $f(z)$ is $q$. If $p_j (j=0,1,2,\dots)$ denotes the number of finite non-zero deficient values of $f^{(j)}(z)$, then we have

$$\sum_{j=0}^{\infty} p_j \le q.$$

In addition, if $f(z)$ is entire, then we have

$$\sum_{j=0}^{\infty} p_j \le \min(\tfrac{q}{2}, 2\lambda).$$

The proof of Theorem 2.2 is based on the following fact. Suppose that $f(z)$ is meromorphic and of order $\lambda(0<\lambda<\infty)$ and that $0$ and $\infty$ are the Borel exceptional values of $f(z)$ in an angular domain $\overline{D}$. If there is a sequence of arcs tending to $\infty$ in $D$, on which $f^{(j)}(z)$ $(j \ge 0)$ approaches sufficiently to a finite non-zero complex value, then there are not any filling disks of order $\lambda$ of $f^{(j)}(z)$ around these arcs.

On the other direction, there exists indeed a meromorphic function of finite order having infinite deficient values, as indicated by Goldberg [18]. Arakelyan [19] proceeded to construct entire functions of order $\lambda(\frac{1}{2}<\lambda<\infty)$ having an infinite set of deficient values. Based on the Arakelyan's method and Mergelyan's theorems, Drasin, Weitsman, Yang and Zhang proved the following theorem [20]:

THEOREM 2.3. Let $\alpha_{jk}$ $(j=0,1,2,\ldots;k=1,2,\ldots,k_j;1\le k_j<\infty)$ be finite complex numbers, with $\alpha_{jk}\ne\alpha_{jk'}(k\ne k')$. Given $\frac{1}{2}<\lambda<\infty$ and an increasing sequence $\{n_j\}$ of integers, there exists an entire function $f(z)$ of order $\lambda$, mean type, such that

$$\delta(\alpha_{jk},f^{(n_j)})>0$$

for all $j$ and $k$.

Let $\Delta_j(f)$ denote $\sum_{\alpha\ne\infty}\delta(\alpha,f^{(j)})$. Fuchs [21] asked if it is possible that $\Delta_j(f)$ is strictly increasing. The above theorem gives an affirmative answer.

## 3. DEFICIENT VALUES AND BOREL DIRECTIONS

In Nevanlinna theory, there is a field named angular distribution theory which investigates the behavior of entire and meromorphic functions for the variable $z$ near some rays. With the inception of Julia direction, Valiron [22] proceeded to prove the existence of Borel direction. Let $f(z)$ be a meromorphic function of order $\lambda$, where $0<\lambda<\infty$. Valiron proved that there exists a ray $\arg z=\theta_0(0\le\theta_0<2\pi)$ with the following property: the equality

$$\overline{\lim_{r\to\infty}}\frac{\log n(r,\theta_0,\varepsilon,f=\alpha)}{\log r}=\lambda$$

holds for any positive number $\varepsilon$ and all the complex values $\alpha$, except at most two values of $\alpha$, where $n(r,\theta_0,\varepsilon,f=\alpha)$ denotes the number of zeros (with their multiplicities) of $f(z)-\alpha$ in the region $(|z|\le r)\cap(|\arg z-\theta_0|\le\varepsilon)$. Such a ray is named a Borel direction (of order $\lambda$) of $f(z)$.

Yang and Zhang [23] obtained an interesting result between numbers of deficient values and Borel directions.

THEOREM 3.1. Let $f(z)$ be a meromorphic function of order $\lambda(0<\lambda<\infty)$. If $p$ is the number of deficient values of $f(z)$ and $q$ is the number of Borel directions of $f(z)$, then we have $p\le q$.

The proof of Theorem 3.1 depends principally on the following lemma.

LEMMA 3.1. Let $f(z)$ be a meromorphic function of order $\lambda(0<\lambda<\infty)$ and have no Borel directions in an angle $\omega_1<\arg z<\omega_2$. Suppose there exists a sequence of positive numbers $R_n$ with $\lim_{n\to\infty}R_n=+\infty$ and a complex number $\alpha_0$ such that for any positive number $\varepsilon$, the measure of the set $E_n$ of values $\omega(\omega_1<\omega<\omega_2)$ with

$$\begin{cases} \log \dfrac{1}{|f(R_n e^{i\omega}) - \alpha_0|} > R_n^{\lambda-\varepsilon} & \text{if} \quad \alpha_0 \neq \infty, \\ \log |f(R_n e^{i\omega})| > R_n^{\lambda-\varepsilon} & \text{if} \quad \alpha_0 = \infty, \end{cases}$$

is larger than $K_1$, which is positive and not dependent on $\varepsilon$, provided that $n$ is sufficiently large. Under these conditions, for a positive number $K_2$ and a sufficiently small positive number $\varepsilon$, we can find a sequence of curves $L_n$ with the following conditions.

1. $L_n$ is located in the region $\omega_1 + 8\alpha \leq \arg z \leq \omega_2 - 8\alpha$, $R_n - 1 \leq |z| \leq R_n$. Its end points are $R_n e^{i(\omega_1 + \alpha_n')}$ and $R_n e^{i(\omega_2 - \alpha_n')}$ $(8\alpha \leq \alpha_n' \leq 9\alpha)$. Denoting by $A_n$ the arc $\{R_n e^{i\omega} | \omega_1 < \omega < \omega_2\}$, the measure of the set of values $\omega$ for which $R_n e^{i\omega}$ belongs to $A_n - L_n$, is less than $K_2$.

2. For every positive number $\eta$, the inequality

$$\begin{cases} \log \dfrac{1}{|f(z) - \alpha_0|} > R_n^{\lambda-\eta} & \text{when} \quad \alpha_0 \neq \infty, \\ \log |f(z)| > R_n^{\lambda-\eta} & \text{when} \quad \alpha_0 = \infty, \end{cases}$$

holds on $L_n$, provided that $n$ is sufficiently large.

Given a meromorphic function $f(z)$ of positive and finite order, one of its deficient values depending on all the zeros of $f(z) - \alpha$ in the plane, is a global concept, whereas the Borel direction is a local concept depending on the behavior of $f(z)$ when $z$ is near this direction. The existence of the deficient value $\alpha$ indicates that $f(z) - \alpha$ has few zeros and a certain part of $f(z)$ varies very slowly. The existence of a Borel direction of $f(z)$ indicates that it assumes most values quite often and varies rapidly in the neighborhood of that direction. Although there are many differences between these two fundamental concepts, it is very interesting that a simple and explicit relation between them has been given by Theorem 3.1.

The estimate $p \leq q$ is best possible. For instance, given any positive integer $n$, there is a meromorphic function of order $\frac{n}{2}$ such that both its numbers of deficient values and of Borel directions are $n$. That is to say $p = q$.

Under the hypotheses of Lemma 3.1, the magnitude of the angle does not exceed $\frac{\pi}{\lambda}$. Using this fact, the conclusion of Theorem 3.1 can be completed as that if $\lambda > \frac{p}{2}$, then we have $p \leq q - 1$ [23].

Let $f(z)$ be a meromorphic function of order $\lambda (0 < \lambda < \infty)$. Its $q (< \infty)$ Borel directions divide the plan into $q$ angles $G_k (k=1,2,\ldots,q)$. Generally speaking, two different deficient values cannot correspond to the same angle

$G_k$, in which $f(z)$ tends rapidly to these two values on a sequence of arcs. If $f(z)$ is entire, then it is impossible that there are two neighboring angles $G_k$, $G_{k+1}(k=1,2,\ldots,q, G_{q+1} = G_1)$, in which $f(z)$ tends rapdily to two different dificient values respectively. More precisely, Yang and Zhang proved the following lemma.

LEMMA 3.2. Let $f(z)$ be an entire function of order $\lambda(0<\lambda<\infty)$ $\lambda(r)$ its precise order and $U(r) = r^{\lambda(r)}$. Suppose that $B_\ell$ : $\arg z = \theta_\ell$ $(\ell = 1,2,3,4; 0 \leq \theta_1 < \theta_2 < \theta_3 < \theta_4 < \theta_1 + 2\pi)$ are four rays and there are no Borel directions in the angles $\Omega_1$: $\theta_1 < \arg z < \theta_2$ and $\Omega_2$: $\theta_3 < \arg z < \theta_4$. If there are two finite distinct values $\alpha_\mu$ $(\mu = 1,2)$, two positive numbers $\eta_\nu$ $(\nu = 1,2)$ and a sequence of positive numbers $R_n$ with $\lim_{n\to\infty} R_n = \infty$ such that

$$\text{Mes } E\{\theta \mid \theta \in \Omega_\mu, \log|f(R_n e^{i\theta}) - \alpha_\mu| < -\eta, U(R_n)\} > \eta_2 \quad (\mu = 1,2)$$

then

$$\theta_3 - \theta_2 \geq \frac{\pi}{\lambda}$$

and

$$(\theta_1 + 2\pi) - \theta_4 \geq \frac{\pi}{\lambda}.$$

Based on Lemma 3.2, Yang and Zhang established

THEOREM 3.2. [23]. Suppose that $f(z)$ is an entire function of order $\lambda$, where $\lambda < \infty$. If $p$ is the number of finite dificient values of $f(z)$ and $q$ is the number of Borel directions of $f(z)$, then we have $p \leq \frac{q}{2}$.

Furthermore, if $q < \infty$, then $p < 2\lambda$.

The estimate $p \leq \frac{q}{2}$ is best possible. For instance, given any positive integer $n$, there exists an entire function of finite positive order such that $p = n$ and $q = 2n$. The function $f(z) = \int_0^z e^{-t^n} dt$ provides such an example. Its order equals $n$. All the finite deficient values of $f(z)$ are $e^{\frac{2k\pi i}{n}} \int_0^\infty e^{-t^n} dt$ $(k=1,2,\ldots,n)$ and all the Borel directions of $f(z)$ are $\arg z = \frac{2k-1}{2n}\pi$ $(k=1,2,\ldots,2n)$. Consequently, $p = \frac{q}{2}$.

The study of many special functions and the Denjoy's conjecture led R. Nevanlinna to the following conjecture.

An entire function of finite order $\lambda(0<\lambda<\infty)$ has $2\lambda$ finite deficient values at most.

In the general case, this conjecture was disproved by Arakelyan [19] in 1966. The second part of Theorem 3.2, however, gives a sufficient condition to make the correctness of the Nevanlinna's conjecture.

The conclusions of Theorem 3.2 were improved by Yang and Zhang themselves.

THEOREM 3.3. [25]. With the hypotheses and notation of Theorem 3.2, we have

$$\sum_{j=-\infty}^{\infty} p_j \leq \frac{q}{2}.$$

Furthermore, if $q < \infty$, then

$$\sum_{j=-\infty} p_j < 2\lambda,$$

where $p_j$ denotes the number of finite, non-zero deficient values of $f^{(j)}(z)$, which is the derivative of order $j$ of $f(z)$, for $j$ positive and the primitive of order $-j$ of $f(z)$ for negative and $f^{(0)}(z)$ the function itself.

THEOREM 3.4. [24]. Suppose that the hypotheses and notations of Theorem 3.2 are conserved and that the plan is divided into $q$ angles $\theta_j < \arg z < \theta_{j+1}$ $(j=1,2,\ldots,q; 0 \leq \theta_1 < \theta_2 < \ldots < \theta_q < \theta_{q+1} = 2\pi + \theta_1)$ by $q$ Borel directions. If, among these $q$ angles, there are $\ell$ angles with the magnitudes larger than $\frac{\pi}{\lambda}$, then we have

$$p \leq \frac{q}{2} - \ell$$

and

$$p < 2\lambda - 2\ell.$$

Zhang [26] proceeded to do research on relations between deficient values, asymptotic values and Julia directions of entire and meromorphic functions. Among others, he proved

THEOREM 3.5. Let $f(z)$ be an entire function of lower order $\mu < \infty$. If $q$ denotes the number of its Julia direction, $\ell$ the number of its different finite asymptotic values, $p$ the number of its finite deficient values and $\ell'$ the number of its finite deficient values, which are simultaneously asymptotic values, then we have $2p - \ell' + \ell \leq q$.

## 4. DISTRIBUTION OF BOREL DIRECTIONS

Concerning the distribution of Borel Directions, a long standing problem is: Given a meromorphic function $f(z)$ of order $\lambda (0 < \lambda < \infty)$, what can we say about the distribution of its Borel Directions?

If $E = \{\theta\}$ is the set of values $\theta$ such that $0 \leq \theta < 2\pi$ and $\arg z = \theta$ are Borel directions of $f(z)$, then according to the Valiron's theorem, $E$ must be a non-empty set. Besides this, it is clear that $E$ is a closed set of real values (mod $2\pi$).

Conversely, Yang and Zhang [27] achieved

THEOREM 4.1. If $\lambda$ is a finite positive number and $E$ is a non-empty set of real values (mod $2\pi$), then we can construct a meromorphic function of order $\lambda$ such that its Borel directions are given precisely by $\{\arg z = \theta | \theta \in E\}$.

Although the proof of Theorem 4.1 is rather long, the idea can be easily understood. It is sufficient to consider the case that $E$ is countable. Choose a sequence of pairs $(a_n, b_n)$ such that $\{a_n\}$ and $\{b_n\}$ are regularly located on the rays $\{\arg z = \theta | \theta \in E\}$ and for every $n$, $a_n$ and $b_n$ are very closed. Set

$$f(z) = \prod_{n=1}^{\infty} \left( \frac{z-a_n}{z-b_n} \right)^{m_n} .$$

where integers $m_n$ are chosen to ensure that the order of $f(z)$ equals $\lambda$.

On every ray $L_1$: $\arg z = \theta$, $\theta \in E$, $f(z)$ varies very rapidly from zero at $a_n$ to pole at $b_n$. From this point of view, $L_1$ can be proved as a Borel direction of $f(z)$. For another ray $L_2$: $\arg z = \theta$, $\theta \overline{\in} E$, it is easy to see that $\frac{z-a_n}{z-b_n}$ is nearly equal to 1 when $z$ is located in a small angle of $L_2$ and $|z|$ is sufficiently large. Thus $L_2$ cannot be a Borel direction of $f(z)$.

A general case can be discussed. Suppose that $f(z)$ has a Borel direction. Let $E$ be the set of real values $\theta$(mod $2\pi$) such that all the rays $\arg z = \theta$ are Borel directions of various order of $f(z)$. Clearly $E$ is non-empty and closed. For any number $\theta$ of $E$, denotes by $\rho(\theta)$ the order of Borel direction $\arg z = \theta$. Thus a function $\rho(\theta)$ is defined on $E$. It is easy to see that $0 \le \rho(\theta) \le \infty$ and $\overline{\lim}_{\substack{\theta\to\theta_0 \\ \theta\in E}} \rho(\theta) \le \rho(\theta_0)$ for any $\theta_0 \in E$. i.e. $\rho(\theta)$ is upper semi-continuous on $E$ and the order of $f(z)$ is equal to $\max_{\theta\in E} \rho(\theta)$. Conversely the following theorem was obtained by Yang and Zhang [27].

THEOREM 4.2. If $E$ is a non-empty and closed set of real values (mod $2\pi$) and $\rho(\theta)$ is an upper semi-continuous function on $E$, $0 \le \rho(\theta) \le \infty$, then there exists a meromorphic function $f(z)$ of order $\max_{\theta\in E} \rho(\theta)$ such that it takes all the rays $\arg z = \theta$, $\theta \in E$ as Borel directions of order $\rho(\theta)$ of $f(z)$ and it has no other Borel directions.

For entire functions, there are some classic results due to Valiron and Cartwright, [22].

Let $f(z)$ be an entire function of finite order $\lambda$. If $\lambda$ is larger than $\frac{1}{2}$, then $f(z)$ has two Borel directions at least.

If $\lambda$ is larger than 1 and $f(z)$ has exactly two Borel directions, then the magnitude of their angle equals $\frac{\pi}{\lambda}$.

Yang and Zhang [28] considered the case of meromorphic functions having a deficient value. They obtained the following theorem.

THEOREM 4.3. Let $f(z)$ be meromorphic of order $\lambda(> \frac{1}{2})$. If $f(z)$ has a deficient value, then there are two Borel directions of $f(z)$ at least. Moreover, the magnitude of their angle does not exceed $\frac{\pi}{\lambda}$.

In the hypotheses of Theorem 4.3, a deficient value of $f(z)$ can be replaced by a deficient value of $f^{(k)}(z)$.

Yang and Zhang [29] investigated also the distribution of Borel directions of entire functions. For instance, they proved:

Let $f(z)$ be an entire function of order $\lambda(0 < \lambda < \infty)$. Denote by $p$ the number of its finite deficient values and $q$ the number of its Borel directions. If $\lambda > p$, then $q \geq 2p + 1$. If $\lambda > p + \frac{1}{2}$, then $q \geq 2p + 2$.

With the same notations of $f(z)$, $\lambda$, $p$ and $q$, if $q = 2p$ and $\lambda = p$, then these Borel directions divide the plane into $q$ equal angles.

## 5. COMMON BOREL DIRECTIONS OF A MEROMORPHIC FUNCTION AND ITS DERIVATIVES

Concerning Borel directions of meromorphic functions, an interesting question is the following: Is it true that a meromorphic function and its derivative always have a common Borel direction? At the end of his pamphlet entitled "Directions de Borel des fonctions méromorphes", Valiron [22] made some remarks on this question and pointed out where the difficulty resides. Rauch [30] gave a sufficient condition in order that a Borel direction of an entire function $f(z)$ is also a Borel direction of $f'(z)$. Several years later, Chuang [31] generalized the theorem of Rauch to the case of meromorphic functions and proved the following.

THEOREM 5.1. Let $\Delta$ be a Borel direction of order $\rho$ of a meromorphic function $f(z)$ of finite positive order $\rho$. If $f(z)$ has two Borel exceptional values, $a, b$ $(a \neq b)$ in an angle containing $\Delta$ in its interior, then $\Delta$ is also a Borel direction of order $\rho$ of the derivative $f'(z)$ of $f(z)$. Moreover if one of the values $a, b$ is $\infty$, then $\Delta$ is also a Borel direction of order $\rho$ of all the successive derivatives $f^{(n)}(z)$ $(n=1,2,\ldots)$ of $f(z)$.

Recently Chuang [32] gave some complements to Theorem 5.1, which may be stated as follows:

Let $f(z)$ be a non-constant meromorphic function of finite order and $0 < \rho < +\infty$ a number. A ray $\Delta : \arg z = \theta_0$ is called a Borel direction of order at least $\rho$ of $f(z)$, if, no matter how small is the number $\eta > 0$, the inequality

$$\overline{\lim_{r\to+\infty}} \frac{\log^+ n(r,\theta_0,\eta,a)}{\log r} \geq \rho \tag{5.1}$$

holds for each value $a$, except at most for two exceptional values, where $n(r,\theta_0,\eta,a)$ denotes the number of roots of the equation $f(z) = a$ in the sector $S: |\arg z - \theta_0| < \eta$, $|z| < r$, each root being counted according to its order of multiplicity. If the conditions of this definition are still satisfied in replacing in the inequality (5.1) $n(r,\theta_0,\eta,a)$ by $n_1(r,\theta_0,\eta,a)$ which denotes the number of simple roots (roots of order 1) of the equation $f(z) = a$ in $S$, we call $\Delta$ a simple Borel direction of order at least $\rho$ of $f(z)$.

THEOREM 5.2. If $\Delta$ is a Borel direction of order at least $\rho$ of the function $f(z)$ and if for a number $0 < \eta_0 < \frac{\pi}{2}$ and two values $a_i (i=1,2; a_1 \neq a_2)$ we have

$$\overline{\lim_{r\to+\infty}} \frac{\log n(r,\theta_0,\eta_0,a_i)}{\log r} < \rho \quad (i=1,2),$$

then $\Delta$ is a simple Borel direction of order at least $\rho$ of $f(z)$. Moreover if one of the values $a_i (i=1,2)$ is $\infty$, then, $(\alpha_0^{(n)}, \alpha_1^{(n)}, \ldots, \alpha_{p_n}^{(n)})$ $(n = 1,2,\ldots)$ being a sequence of systems of integers such that

$$\alpha_0^{(1)} = 0, \quad \sum_{j=0}^{p_n} \alpha_j^{(n)} \neq 0 \quad (n = 1,2,\ldots),$$

$\Delta$ is a simple Borel direction of order at least $\rho$ of each of the functions $f_n(z)$ $(n = 1,2,\ldots)$ defined by the relations:

$$f_n(z) = \prod_{j=0}^{p_n} \{f_{n-1}^{(j)}(z)\}^{\alpha_j^{(n)}} \quad (n = 1,2,\ldots; f_0 = f).$$

In particular, if we take $p_n = 1$, $\alpha_0^{(n)} = 0$, $\alpha_1^{(n)} = 1$ $(n = 1,2,\ldots)$, we have $f_n(z) = f^{(n)}(z)$ $(n = 1,2,\ldots)$.

By a different method, Zhang [33], showed that the condition in Theorem 5.1 can be replaced by a weaker one; it is sufficient to suppose that $f(z)$ has one finite Borel exceptional value in an angle containing $\Delta$ in its interior. Among other results, he also gave a generalization to the case of meromorphic functions of the following important theorem of Milloux [34]:

Every Borel direction of order $\rho$ of the derivative $f'(z)$ of an entire function $f(z)$ of finite positive order $\rho$ is a Borel direction of order $\rho$ of $f(z)$.

Milloux's proof is very long and complicated. Now the following theorem, from which one can deduce Milloux's Theorem immediately, was simply proved by Yang [35]:

THEOREM 5.3. Suppose that $f(z)$ is meromorphic and of order $\lambda (0 < \lambda < \infty)$ and that $\infty$ is a Borel exceptional value of $f(z)$ in the angle $|\arg z-\theta_0| < \tau_0 (0 \le \theta_0 < 2\pi, \tau_0 > 0)$. Let

$$\Gamma_n: |z-z_n| < \varepsilon_n |z_n|, \quad |z_{n+1}| > 2|z_n|,$$

$$\lim_{n\to\infty} \arg z_n = \theta_n, \quad \lim_{n\to\infty} \varepsilon_n = 0.$$

be a sequence of disks such that $f'(z)$ takes every complex value at least $R_n^{\lambda-\varepsilon'_n}$ times in $\Gamma_n$ except at most some values enclosed in two small disks with the radius $\delta_n$ on Riemann sphere, where $\lim_{n\to\infty} \varepsilon'_n = \lim_{n\to\infty} \delta_n = 0$. If $G_n$ denotes the regions

$$\left( \frac{R_N^{1-\eta_n}}{2} < |z| < 2R_n^{1+\eta_n} \right) \cap (|\arg z| < 20\pi\eta_n)$$

with

$$\eta_n = 4\pi \varepsilon_n^{*1/2} \to 0 \quad (n\to\infty),$$

where

$$\varepsilon_n^* = \max\left( \varepsilon_n + \arg z_n, \frac{2\varepsilon'_n}{\lambda}, \frac{2\beta_n}{\lambda}, \frac{1}{(\log|z_n|)^{1/2}} \right),$$

$$\rho_n = \max_{|z_n|^{1/2} \le r \le |z_n|} \left( \frac{\log T(r,f)}{\log r} - \lambda \right)$$

then $(G_n)$ must contain a subsequence $(G_{n_k})$ such that $f(z)$ takes every complex value at least $R_{n_k}^{\lambda-\varepsilon''_{n_k}}$ times in $G_{n_k}$, except some values enclosed in two small spherical disks with the radius $\delta'_{n_k}$, where $\lim_{k\to\infty} \varepsilon''_{n_k} = \lim_{k\to\infty} \delta'_{n_k} = 0$

Quite recently, Yang Lo and Zhang Qingde [36] obtained the following result.

THEOREM 5.4. If $f(z)$ is meromorphic and of order $\lambda (0 < \lambda < \infty)$, then there exists a direction $\arg z = \theta_0 (0 \le \theta_0 < 2\pi)$ such that, for every positive integer $k$, any positive number $\varepsilon$ and two arbitrary finite complex values $\alpha$ and $\beta$ $(\beta \ne 0)$, the equality

$$\overline{\lim_{r\to\infty}} \frac{\log\{n(r,\theta_0,\varepsilon,f=\alpha)+n(r,\theta_0,\varepsilon,f^{(k)}=\beta)\}}{\log r} = \lambda$$

always holds.

If such directions are named as directions of Borel type, then it is interesting that there exists a meromorphic function of finite and non-zero order such that one of its directions of Borel type is not its Borel direction.

## 6. EXTENSIONS OF MIRANDA THEOREM ON NORMAL FAMILIES

For a long time, Montel had a presentiment of the exactitude of the following theorem:

Let $F$ be a family of holomorphic functions $f(z)$ in a domain $D$. If in $D$ each function $f(z)$ of $F$ does not take the value $0$ and its derivative of order $k(k \geq 1$, fixed) $f^{(k)}(z)$ does not take the value $1$, then $F$ is normal in $D$.

Under the impulsion of Montel, Bureau [37] utilizing a method of Nevanlinna, obtained some results which make the above theorem very likely to be true. Then in completing skillfully the method of Bureau, Miranda [38] proved the above theorem. Later on, Valiron [39] obtained more general results by a different method. One of them is a generalization of the Miranda theorem to the case where $f^{(k)}(z)$ is replaced by a linear combination

$$a_0 f(z) + a_1 f'(z) + \ldots + a_k f^{(k)}(z)$$

with constant coefficients $a_j$ $(j=0,1,2,\ldots,k;\ a_k \neq 0)$. This work of Valiron was succeeded by Chuang [40] [41]. First he proved a general theorem on holomorphic functions in the unit disk. Then on the basis of it, he obtained various generalizations of Schottky theorem and criteria of quasi-normality for families of holomorphic functions. However here we mention only the following theorem which he proved by the method of Miranda:

THEOREM 6.1. Let $F$ be a family of holomorphic functions $f(z)$ in a domain $D$. If there are holomorphic functions $a_j(z)$ $(j=0,1,2,\ldots,k)$, $u(z)$ and $v(z)$ in $D$ such that

$$a_k(z) \not\equiv 0, \quad \sum_{j=0}^{k} a_j(z)u^{(j)}(z) \not\equiv v(z)$$

and that for each function $f(z)$ of $F$, the equations

$$f(z) = u(z), \quad \sum_{j=0}^{k} a_j(z)f^{(j)}(z) = v(z)$$

have no root in $D$, then the family $F$ is normal in $D$.

After the preceding works, Milloux [42] [43] gives an extension of the second fundamental theorem of Nevanlinna to the following form:

$$T(r,f) < (k+1)N(r,f) + N(r,\frac{1}{f}) + N(r,\frac{1}{\psi-1}) + S(r),$$

where

$$\psi(z) = \sum_{j=0}^{k} a_j(z)f^{(j)}(z),$$

$a_j(z)(j=0,1,2,\ldots,k)$ being holomorphic functions satisfying certain conditions. He also obtained, by means of a suitable auxiliary function, the inequality

$$T(r,f) < (2k+1)N(r,\frac{1}{f}) + (k+1)N(r,\frac{1}{f-1}) + N(r,\frac{1}{\psi-1}) + S(r),$$

which does not contain the counting function $N(r,f)$ for poles.

In 1959, Hayman [44] proved the following remarkable theorem, in which only two counting functions need to bound the characteristic function.

Let $f(z)$ be meromorphic and $k$ be a positive integer. Then

$$T(r,f) < (2+\frac{1}{k})N(r,\frac{1}{f}) + (2+\frac{2}{k})N(r, \frac{1}{f^{(k)}-1}) + S(r,f).$$

Consequently, if $f(z) \neq 0$ and $f^{(k)}(z) \neq 1$, then $f(z)$ must reduce to a constant. On the basis of this inequality, Hayman asked whether there is a corresponding criterion for normality. The difficulty resided in eliminating the complicated initial values in $S(r,f)$ of the Hayman's inequality. In 1979, Ku was successful in proving [45].

THEOREM 6.2. Let $\mathcal{F}$ be a family of meromorphic functions in a region $D$ and $k$ be a positive integer. If, for every function $f(z)$ of $\mathcal{F}$, $f(z) \neq 0$ and $f^{(k)}(z) \neq 1$, then $\mathcal{F}$ is normal in $D$.

Ku's proof is ingenious. In order to make it natural and direct, the following theorem has been introduced [46] [47].

THEOREM 6.3. If $k$ is a positive integer, $f(z)$ is meromorphic in $|z| < 1$ and $f(z) \neq 0$, $f^{(k)}(z) \neq 1$ there, then either $|f(z)| < 1$ or $|f(z)| > C$ uniformly in $|z| < \frac{1}{32}$, where $C$ is a positive constant which depends only on $k$.

The proof of Theorem 6.3 includes two mutually exclusive cases. The key point of the principal case is that a point $z_0$ can be found such that

$$|f(z_0)| < \frac{1}{2}, \quad |f^{(k)}(z_0)| < \frac{1}{2}, \quad \frac{1}{12} < |f^{(k+1)}(z_0)| < \frac{1}{2},$$

and

$$|f^{(k+2)}(z_0)| \geq 1.$$

Thus the initial values can be eliminated.

In fundamental inequalities which are used to derive criteria for normality, multiple values with high multiplicities usually can be estimated. Thus some precise results were already obtained. For instance, Yang and Zhang [48] proved the following theorem.

THEOREM 6.4. Let $\mathcal{F}$ be a family of holomorphic functions in a region $D$. If, for every function $f(z)$ of $\mathcal{F}$, all the zeros of $f(z)$ have multiplicity $\geq m$ and those of $f^{(k)}(z)-1$ ($k$ is a non-negative integer. $f^{(0)} \equiv f$) have multiplicity $\geq n$, where $\frac{k+1}{m}+\frac{1}{n}<1$, then $\mathcal{F}$ is normal in $D$.

As a consequence, if for every function $f(z)$ of $\mathcal{F}$, $f'(z)f(z)^k \neq 1 (k>1)$, then $\mathcal{F}$ is normal in $D$. Thus Yang and Zhang settled a Hayman's problem in the case of $k>1$. When $k=1$, the corresponding criterion for normality was proved by Oshkin [49a] in 1982.

Yang and Zhang [49] investigated also the normality of families of meromorphic functions. Based on it, Ku [50] obtained the following satisfactory theorem.

THEOREM 6.5. Let $\mathcal{F}$ be a family of meromorphic functions in a region $D$, $a_i (i=1,2,\ldots,p)$ $p(\geq 1)$ distinct finite complex values and $b_j (j=1,2,\ldots,q)$ $q(\geq 1)$ distinct finite non-zero complex values. Suppose that all the poles of $f(z)$ have multiplicities $\geq s(\geq 1)$, all the zeros of $f(z)-a_i$ have multiplicities $\geq m_i(\geq 2)$ and those of $f^{(k)}(z)-b_j$ have multiplicities $\geq n_j(\geq 2)$. If

$$\sum_{i=1}^{p}\frac{kq+1}{m_i}+\sum_{j=1}^{q}\frac{1}{n_j}+\frac{1}{s}\left(1+k\sum_{j=1}^{q}\frac{1}{n_j}\right)<pq,$$

$$q>\frac{kq+1}{m_i} \quad (i=1,2,\ldots,p),$$

then $\mathcal{F}$ is normal in $D$.

As a consequence, Ku settled another problem of Hayman which says: If for every meromorphic function $f(z)$ of a family $\mathcal{F}$ in $D$, $f'(z)\{f(z)\}^k \neq 1$ with $k \geq 3$, then $\mathcal{F}$ is normal.

Based on [51], Li [51a] recently gave a proof to the following criterion which was posed by Hayman.

THEOREM 6.6. Let $\mathcal{F}$ be a family of meromorphic functions in a region $D$, $k(\geq 5)$ an integer and $\alpha$ a finite non-zero complex value. If $f'(z)-\alpha f(z)^k \neq 1$ for every function $f(z)$ of $\mathcal{F}$, then $\mathcal{F}$ is normal in $D$.

## 7. ASYMPTOTIC VALUES AND ASYMPTOTIC PATHS

Let $f(z)$ be an entire function. We say that a value $a$ is an asymptotic value of $f(z)$, if there is a path $\Gamma$ extending to $\infty$, such that as $z \to \infty$ along $\Gamma$, $f(z) \to a$. The path $\Gamma$ is called an asymptotic path corresponding to the value $a$. By a theorem of Iversen [52], $\infty$ is an asymptotic value of every non-constant entire function. On the other hand, Ahlfors [53] proved the following conjecture of Denjoy: An entire function of finite order $\rho$ can have at most $2\rho$ distinct finite asymptotic values.

Concerning Iversen's theorem, Boas [54] obtained the following result: Let $f(z)$ be a transcendental entire function. Then there exists a path $\Gamma$ extending to $\infty$, such that for every $n$, we have

$$\left|\frac{f(z)}{z^n}\right| \to +\infty \quad \text{as} \quad z \to \infty \quad \text{along} \quad \Gamma.$$

On the other hand, Huber [55] proved that for every $\lambda > 0$ there exists a path $\Gamma_\lambda$ extending to $\infty$ such that

$$\int_{\Gamma_\lambda} |f(z)|^{-\lambda} |dz| < +\infty,$$

provided that $f(z)$ does not reduce to a polynomial. Zhang [56] proved the following theorems which improve the above results of Boas and Huber.

THEOREM 7.1. Let $f(z)$ be an entire function and let $\nu(r)$ be an increasing function of $r$ $(0 < r < +\infty)$ satisfying the following conditions:

1) $\lim\limits_{r\to+\infty} \nu(r) = +\infty$,

2) $\overline{\lim\limits_{r\to+\infty}} \dfrac{\log \nu(r)}{\log r} = 0$

3) $\overline{\lim\limits_{r\to+\infty}} \dfrac{\nu(r^{1+\varepsilon})}{\nu(r)} = C(\varepsilon) < +\infty \quad (\varepsilon > 0)$,

where $C(\varepsilon) \geq 1$, $C(\varepsilon) \to 1$ as $\varepsilon \to 0$. Suppose that

$$\underline{\lim}_{r\to+\infty} \frac{\log M(r,f)}{\nu(r)} = +\infty .$$

Then there exists a path $\Gamma$ extending to $\infty$, such that

$$\underline{\lim}_{\substack{|z|\to+\infty \\ z\in\Gamma}} \frac{\log|f(z)|}{\nu(|z|)} = +\infty.$$

THEOREM 7.2. Suppose that $f(z)$ is a trancendental entire function of finite order, then there exists a path $\Gamma$ extending to $\infty$, such that for every $\lambda > 0$,

$$\int_\Gamma |f(z)|^{-\lambda} |dz| < +\infty.$$

He [57] proved the following theorems on asymptotic paths.

THEOREM 7.3. Let $f(z)$ be a transcendental entire function of finite order $\rho$. Then there exists a rectifiable path $\Gamma$ extending to $\infty$, such that

$$\lim_{\substack{|z|\to+\infty \\ z\in\Gamma}} \frac{\log|f(z)|}{\log|z|} = +\infty$$

and that, $\ell(r)$ being the length of the part of $\Gamma$ in the circle $|z| \le r$, we have

$$\overline{\lim_{r\to+\infty}} \frac{\ell(r)}{r^{1+\frac{\rho}{2}+\varepsilon}} = 0$$

where $\varepsilon > 0$. Moreover if $f(z)$ has a finite asymptotic value $a$, then there exists a rectifiable path $\Gamma$ extending to $\infty$, such that

$$\lim_{\substack{|z|\to+\infty \\ z\in\Gamma}} f(z) = a$$

and

$$\ell(r) = O(r\sqrt{U(r)}),$$

where $U(r) = r^{\rho(r)}$ and $\rho(r)$ is a precise order of $f(z)$.

Concerning Ahlfors theorem mentioned above, Ahlfors and other authors have studied the class of entire functions for which the number of distinct finite asymptotic values attains the limit $2\rho$. These works were succeeded by Zhang [58] [59] whose results may be summarized as follows:

THEOREM 7.4 Let $f(z)$ be a transcendental entire function of finite order $\rho$. Assume that $f(z)$ has $k (k \ge 1)$ distinct finite asymptotic values $a_i (i=1,2,\ldots,k)$ with corresponding asymptotic paths $\Gamma_i (i=1,2,\ldots,k)$ which are supposed to be simple curves issued from the point $z = 0$ and having no other common to each other. If $k = 2\rho$, then the following assertions hold:

1) $f(z)$ has no finite deficient value.

2) For any number $\theta (0 \le \theta < 2\pi)$, the ray $\arg z = \theta$ is either a Julia direction of $f(z)$ or is such that

$$\lim_{\substack{|z|\to+\infty \\ \arg z=\theta}} \frac{\log\log|f(z)|}{\log|z|} = \rho.$$

3) $D_i$ being the domain bounded by the adjacent paths $\Gamma_i$ and $\Gamma_{i+1}$ $(1 \le i \le k,\ L_{k+1} = L_1)$, there exists in $D_i$ a path $L_i$ extending to $\infty$, such that

$$\lim_{\substack{|z|\to+\infty \\ z\in L_i}} \frac{\log \log |f(z)|}{\log |z|} = \rho.$$

4) The angle between any two adjacent Borel directions of order $\rho$ of $f(z)$ does note exceed $\pi/\rho$.

5) If the number of Borel directions of order $\rho$ of $f(z)$ is finite, the paths $\Gamma_i (i = 1,2,\dots,k)$ are respectively asymptotic to $k$ Borel directions $\Delta_i$ $(i = 1,2,\dots,k)$ of order $\rho$ of $f(z)$, the angle between any two adjacent directions $\Delta_i$ and $\Delta_{i+1}$ being equal to $\pi/p$. To say that $\Gamma_i$ is asymptotic to $\Delta_i : \arg z = \theta_i$ means that to each $\varepsilon > 0$ corresponds $R$ such that the part of $\Gamma$ exterior to the disk $|z| \le R$ belongs to the domain $|\arg z - \theta| < \varepsilon$, $R < |z| < +\infty$.

## 8. MEROMORPHIC FUNCTIONS IN THE UNIT DISK OR IN AN ANGLE

Let $f(z)$ be a meromorphic function of order $\lambda (0 < \lambda < \infty)$ in the unit disk. Valiron [60] established the existence of Borel point of order $\lambda + 1$, i.e. there exists a point $z_0 = e^{i\phi_0}$ such that if $G(z_0,\varepsilon)$ denotes the region, $(|z| < 1) \cap (|z-z_0| < \varepsilon)$, then the series

$$\sum \{1-r_n(G(z_0,\varepsilon), f = \alpha)\}^{\tau}$$

converges for $\tau > \lambda + 1$ and every complex value $\alpha$, provided that $\varepsilon$ is sufficiently small and diverges for $\tau < \lambda + 1$ and every small positive number $\varepsilon$, except for two values of $\alpha$ at most.

It is natural to ask whether there exists a common Borel point of order $\lambda + 1$, to $f(z)$ and its derivative. Yang and Shiao proved [61]

THEOREM 8.1. Let $f(z)$ be a meromorphic function of finite order in the unit disk. If its derivative $f'(z)$ takes $z_0 = e^{i\phi_0}$ as a Borel point of order $\tau (>1)$, then $z_0$ is also a Borel point of order $\ge\tau$ of $f(z)$.

As a consequence, every meromorphic function $f(z)$ of order $\lambda (0 < \lambda < \infty)$ in the unit disk, has a Borel point of order $\lambda + 1$ with its successive derivatives.

For meromorphic functions growing slowly in the unit disk, Xing [62] proved the following conjecture posed by Milloux.

THEOREM 8.2. Let $f(z)$ be a meromorphic function in the unit disk and satisfy

$$\overline{\lim_{r\to 1}} \frac{T(r,f)}{\log \frac{1}{1-r}} = \infty ,$$

then there exists a series of filling disks in which the sight of the angles to the circumference of the unit disk tends to zero.

Now suppose that $f(z)$ is holomorphic in an angle

$$S = S(\alpha,\beta) = \{z : \alpha \le \arg z \le \beta,\ |z| \ge 0\}.$$

Write

$$S' = S(\alpha',\beta'), \quad \text{where} \quad \alpha < \alpha' < \beta' < \beta,$$

and

$$M(r,S) = \max_{\alpha\le\theta\le\rho} |f(re^{i\theta})|.$$

Let $n(r,a,S)$ be the number of zeros of $f(z) - a$ in the sector

$$S(r) = \{z : z \in S,\ |z| \le r\}.$$

The order of $f$ in $S$ is defined by

$$K(S) = \overline{\lim_{r\to\infty}} \frac{\log^+ \log^+ M(r,S)}{\log r}.$$

The order of the a-points of $f$ is defined by

$$K(a,S) = \overline{\lim_{r\to\infty}} \frac{\log n(r,a,S)}{\log r}.$$

The following result is classical and due to Nevanlinna.

If for some $S'$, $K(S') > \frac{\pi}{\rho - \alpha'}$, then $K(a,S) \ge K(S')$ holds for every finite complex value $a$ with one exception at most.

Littlewood put forward the conjecture which says: If $K(S') \ge \lambda$ and $K(0,S') \ge \lambda$ for some positive $\lambda$, then $K(a,S) \ge \lambda$ for every finite $a$ with at most one exception or at least for most values.

Recently Hayman and Yang [63] proved the following theorem.

THEOREM 8.3. Suppose that $r_\nu$ is an increasing sequence of positive numbers such that

$$\lim_{\nu\to\infty} r_\nu = \infty$$

and

$$\lim_{\nu\to\infty} \frac{\log r_{\nu+1}}{\log r_\nu} = 1. \tag{8.1}$$

Suppose further that

$$\underline{\lim_{\nu\to\infty}} \frac{\log\log M(r_\nu, S')}{\log r_\nu} \geq \lambda > 0.$$

If $K(0,S') \geq \lambda$, then $K(a,S) \geq \lambda$ for every finite complex value $a$ with one exception at most.

The difference between Littlewood's conjecture and Theorem 8.3 resides in the condition (8.1). Unfortunately, this condition is essential.

THEOREM 8.4. Suppose that $\delta$, $\eta$ and $\lambda$ are given such that

$$0 < \delta < \frac{\pi}{2}, \quad 1 < \eta < 2, \quad \frac{2}{3} < \lambda < 1,$$

and $r_\nu$ is a sequence tending to $\infty$ with $\nu$. There exists an entire function $f(z)$ of order 1, mean type, such that if $S = S(-\frac{\pi}{2}+\frac{3}{4}\delta, \frac{\pi}{2}-\frac{3}{4}\delta)$ and $S' = S(-\frac{\pi}{2}+\delta, \frac{\pi}{3}-\delta)$, then $f(z)$ has order $\lambda$ in $S$ and

$$\underline{\lim} \frac{\log|f(z)|}{|z|^\lambda} > 0$$

as $z \to \infty$ in $S$ outside the sequence of annuli

$$r_\nu < |z| < r_\nu^\eta \qquad \text{for } \nu = 1, 2, \ldots .$$

Further, $K(0,S') = \lambda$ but for $a \neq 0$,

$$K(a,S) \leq \lambda' = \lambda - \frac{(1-\lambda)(\eta-1)}{S} .$$

Finally we mention a result of Lai [64] who succeeded in finding out the precise value of Hayman constant $A$ in Landau theorem. He proved the following theorem:

THEOREM 8.5. If the function

$$f(z) = a_0 + a_n z^n + a_{n+1} z^{n+1} + \ldots, \quad a_n \neq 0$$

is holomorphic for $|z| < R$ and does not take the values 0 and 1, then

$$R^n \leq 2\frac{|a_0|}{|a_n|}\{|\log|a_0||+A\}, \ A = \frac{\Gamma(\frac{1}{4})^4}{4\pi^2} = 4.37\ldots .$$

## 9. ALGEBROIDAL FUNCTIONS

In China, research on algebroidal functions started from 1955, when Zhao [65] discussed the inverse function of entire algebroidal function with two branches, and then Hiong [66] [67] extended a theorem of Milloux on meromorphic functions to the case of algebroidal functions. These works were followed by those of He, Lü and Gu. We are going to give an account of some of their results.

In what follows, by a $\nu$-valued algebroidal function, we mean a function $w(z)$ defined by an irreducible equation

$$A_\nu(z)w^\nu + A_{\nu-1}(z)w^{\nu-1} + \ldots + A_0(z) = 0$$

where $A_j(z)\,(j=0,1,\ldots)$ are entire functions without common zero.

He [68] [69] obtained some extensions to algebroidal function of the second fundamental theorem and applied them to the study of the deficient values, multiple values and uniqueness problem of algebroidal functions. One of his results is the following $\gamma\,(\geq 1)$ being an integer, denote by $\overline{N}^{\gamma)}(r,a)$ the counting function of the zeros of $w(z) - a$, where each zero of order $\tau$ is counted once, if $\tau \leq \gamma$, and neglected, if $\tau > \gamma$. Then we have the inequality

$$\{\gamma p - 2\nu(\gamma+1)\}T(r,w) < \gamma \sum_{i=1}^{p} \overline{N}^{\gamma)}(r,a_i) + S_p(r,w). \tag{9.1}$$

Using (9.1), He proved

THEOREM 9.1. Let $w(z)$ be a $\nu$-valued algebroidal function. Denote by $\overline{E}^{\gamma)}(a,w)$ the set of those $a$-points of $w(z)$, whose order of multiplicity $\leq\gamma$, each of them being counted once. Then $w(z)$ is uniquely determined by $p = 4\nu + 1 + [\frac{2\nu}{\gamma}]$ sets $\overline{E}^{\gamma)}(a_i,w)$ $(i = 1,2,\ldots,p)$. In particular, if $\gamma = 1,2,\ldots,\nu$ and $2\gamma+1$, then $p = 6\nu + 1$, $5\nu + 1$, $4\nu + 3$ and $4\nu + 1$ respectively.

Lü [71] studied the relationship between the order of an algebroidal function and the number of the direct transcendental singularities of its inverse function and proved

THEOREM 9.2. Let $w(z)$ be a $\nu$-valued algebroidal function and denote by $\lambda$ and $p$ its lower order and the number of the direct transcendental singularities of its inverse function respectively. Then

$$p \leq \begin{cases} 2\nu\lambda + 2(\nu-1), & \text{if } \lambda \geq \frac{1}{2} \\ 2\nu, & \text{if } \lambda < \frac{1}{2\nu}. \end{cases}$$

In view of the existence of Borel directions of meromorphic functions, it is natural to ask: for a $\nu$-valued algebroidal function $w(z)$ of finite positive order $\rho$, does there exist a direction $\arg z = \theta_0$ $(0 \le \theta_0 < 2\pi)$ such that, no matter how small it is the positive number $\varepsilon$, we have

$$\overline{\lim_{r\to+\infty}} \frac{\log n(r,\theta_0,\varepsilon,w=a)}{\log r} = \rho$$

for every value $a$, except a "small" set $E$ of values $a$.

Rauch [72] proved that, for entire algebroidal functions, this is true, when $E$ is a set of linear measure zero. Then Toda [73] showed that $E$ is a countable set. Toda's result was improved by Lü [74] who proved that $E$ contains at most $8\nu - 6$ values. Finally Lü and Gu [75] obtained the expected result that $E$ contains at most $2\nu$ values.

Ozawa [76] proved that the lower order of a $\nu$-valued entire algebroidal function with $\nu$ finite deficient values is positive. He conjectured that for $\nu$-valued meromorphic algebroidal function with $\nu + 1$ deficient values, the lower order is also positive. Gu [77] gave an affirmative answer to this conjecture.

## 10. GENERALIZATIONS OF MALMQUIST THEOREM

Malmquist had investigated the problem of the existence of meromorphic or algebroidal solution of first ordinary differential equations and proved a remarkable theorem. In 1933, Yosida gave an elegant proof and extensions of Malmquist theorem by using Nevanlinna's value distribution theory. Since then Nevanlinna's theory has become an important tool for the study of the global properties of ordinary differential equations in complex domain.

In their works [79] ~ [82], He and Xiao studied the meromorphic or algebroidal solution of generalized algebraic differential equations and obtained some theorems of the Malmquist type. One of their results may be stated as follows.

Consider an equation of the form:

$$\Omega(z,w) = R(z,w), \tag{10.1}$$

where

$$\Omega(z,w) = \sum a_{i_0\ldots i_n}(z)(w)^{i_0}\ldots(w^{(n)})^{i_n}$$

is a differential polynomial, and

$$R(z,w) \equiv P(z,w)/Q(z,w) \equiv \sum_{k=0}^{p} a_k(z)w^k / \sum_{j=0}^{a} b_j(z)w^j,$$

the coefficients $\{a_{i_0 \dots i_n}(z)\}$, $\{a_k(z)\}$ and $\{b_j(z)\}$ being meromorphic functions.

By definition, an algebroidal solution $w(z)$ of equation (10.1) is said to be admissible, if $w(z)$ satisfies

$$\sum T(r, a_{i_0 \dots i_n}) + \sum T(r, a_k) + \sum T(r, b_j) = o\{T(r,w)\}$$

outside a possible exceptional set of values of $r$ of finite linear measure.

He and Xiao proved [82]

THEOREM 10.1. If the equation (10.1) admits at least an admissible algebroidal solution with $\nu$ branches, then

$$q \le 2\sigma(\nu=1) \quad \text{and} \quad p \le q + \lambda + \bar{\mu}\nu(1-\theta(w,\infty)), \tag{10.2}$$

where $\sigma = \mathrm{Max}\{\sum_{\alpha=1}^{n} (2\alpha-1) i_\alpha\}$, $\lambda = \mathrm{Max}\{\sum_{\alpha=0}^{n} i_\alpha\}$, $\bar{\mu} = \mathrm{Max}\{\sum_{\alpha=1}^{n} \alpha i_\alpha\}$ and

$$\theta(w,\infty) = 1 - \overline{\lim_{r\to+\infty}} \frac{\bar{N}(r,w)}{T(r,w)}.$$

The following two examples show that in (10.2) the upper bounds of $q$ and $p$ can be attained.

EXAMPLE 1. The 2-valued algebroidal function $w(z)$ defined by

$$w^2 + (\mathrm{tg} z)w + z^2 = 0$$

is an admissible solution of the equation

$$\frac{dw}{dz} = \frac{w^4 + (1+2z^2)w^2 + 2zw + z^4}{z^2 - w^2}.$$

We have $\nu = q = 2$, $\sigma = \bar{\mu} = 1$, $2\sigma(\nu-1) = 2$.

EXAMPLE 2. The $\nu$-valued algebroidal function $w(z)$ defined by

$$(\cos z)w^\nu - (1+z^2)\sin z = 0$$

is an admissible solution of the equation

$$\frac{dw}{dz} = \frac{w^{2\nu} - 2zw^\nu - (1+z^2)^2}{\nu(1+z^2)w^{\nu-1}}.$$

We have $p = 2$, $q = v - 1$, $\lambda = \bar{\mu} = \sigma = 1$, $\theta(w,\infty) = 0$, $q + \lambda + \bar{\mu}\nu = 2\nu$.

## 11. ON THE GROWTH OF FUNCTIONS

### A. Borel's Theorem and Type-functions

In his fundamental paper on the zeros of entire functions [84], Borel studied the comparative growth of two functions of very rapid growth and discovered the following theorem:

Let $f(x)$ be an unbounded continuous nondecreasing function for $x \geq 0$ and $\alpha$, $\beta$ two positive numbers. Then the inequality

$$f(X) < f(x)^{1+\beta}, \quad X = x + \frac{1}{\{\log f(x)\}^{\alpha}}$$

holds, except at most for values of $x$ which can be enclosed in a sequence of intervals of finite total length.

This theorem has had much influence on the subsequent development of the theory of entire or meromorphic functions. Thus in his theory of meromorphic functions [1], Nevanlinna applied this theorem to the estimation of the error term $S(r)$ of the fundamental inequality

$$(q-2)T(r,f) < \sum_{\nu=1}^{q} N(r,a_{\nu}) - N_1(r) + S(r)$$

and obtained the important result

$$S(r) = O(\log T(r,f) + \log r),$$

in excluding possibly a sequence of intervals of finite total length, if the meromorphic function $f(z)$ is of infinite order.

On the other hand, inspired by Borel's theorem, Blumenthal [85] introduced the notion of type-functions in his work on entire functions of infinite order. A type-function may be defined to be an unbounded non-decreasing function $f(z)$ defined in an interval $x \geq x_0$, whose growth satisfies a condition of regularity. For example, in Borel's theorem, if the inequality holds for all sufficiently large values of $x$, then the function $f(x)$ is a type-function. The work of Blumenthal was followed by that of Hiong [86] on meromorphic functions of infinite order. Of such a function $f(z)$, he gave an appropriate definition of order which, instead of the quantity $\rho = +\infty$, is an unbounded non-decreasing function $\rho(r)$ $(r \geq r_0)$ such that

$$\overline{\lim_{r\to+\infty}} \frac{\log T(r,f)}{\rho(r)\log r} = 1$$

and that the function $U(r) = r^{\rho(r)}$ satisfies the condition

$$\lim_{r\to+\infty} \frac{\log U(r')}{\log U(r)} = 1, \; r' = r\{1 + \frac{1}{\log U(r)}\}.$$

This condition shows that $U(r)$ is a type-function. He proved by a long argument that such a function $\rho(r)$ exists and called every such function $\rho(r)$ an order of the meromorphic function $f(z)$ of infinite order. This defintion of order improves one given by Blumenthal. On the other hand, Valiron [87], introduced the important notion of precise order of an entire or meromorphic function of finite positive order. $f(z)$ being a meromorphic function of finite positive order $\rho$, a precise order of $f(z)$ is, by definition, a function $\rho(r)$ such that

$$\overline{\lim_{r\to+\infty}} \frac{T(r,f)}{r^{\rho(r)}} = 1$$

$$\lim_{r\to+\infty} \rho(r) = \rho, \qquad \lim_{r\to+\infty} \rho'(r)\, r \log r = 0.$$

The latter two relations imply that the function $U(r) = r^{\rho(r)}$ satisfies the condition

$$\lim_{r\to+\infty} \frac{U(kr)}{U(r)} = k^{\rho} \qquad (k > 0).$$

Here again $U(r)$ is a type-function. Valiron proved the existence of such a function $\rho(r)$.

The notion of type-functions plays an important role in the theory of entire or meromorphic functions. Besides it is unnecessary to formulate the definition of type-functions in the restricted form given above, the essential point is that the variation of the function should satisfy a condition of regularity.

B. A Precise Form of Borel's Theorem and Theorems on Type-functions

Borel's theorem was put by Nevanlinna [88] into the following precise form:

Let $u(r)$ be a positive non-decreasing function for $r > 0$, $\phi(t)$ a positive decreasing function for $t > 0$ such that $\int_t^{\infty} \phi(t)dt$ converges, and $r_0$ a value of $r$. Then the total length of the intervals in which

$$u(r+\phi\{u(r)\}) \geq u(r) + 1 \quad (r > r_0),$$

does not exceed the limit $\phi(t_0) + \int_{t_0}^{\infty} \phi(t)dt$ where $t_0 = u(r_0)$.

This theorem of Nevanlinna was further improved by Chuang [89] in replacing there the constant 1 by a function satisfying suitable conditions as follows:

THEOREM 11.1. Let $f(z)$, $\gamma(t)$ and $\delta(x)$ be three functions satisfying the following conditions:

$f(x)$ is continuous, non-decreasing and $f(x) \geq 0$ for $x \geq 0$.

$\gamma(t)$ is positive, continuous and non-increasing for $t \geq 0$; $\int_0^\infty \gamma(t)dt$ converges.

$\delta(x)$ is positive, continuous and non-increasing for $x \geq 0$; $\int_0^\infty \delta(x)dx$ diverges.

Then we have

$$f(x+\gamma\{f(x)\}) - f(x) < \delta(x),$$

except at most in a sequence of intervals in which the total variation of $\int_0^x \delta(x)dx$ does not exceed $\delta(0)\gamma(0) + \int_0^\infty \gamma(t)dt$.

Nevanlinna's theorem corresponds to the particular case $\delta(z) = 1$. Besides this theorem, Chuang proved in the same work a corresponding theorem on type-functions, for the statement of which, we first give two definitions.

DEFINITION 11.1. Let $\gamma(t)$ and $\delta(x)$ be two functions satisfying the conditions in Theorem 1. A function $f(x)$ is said to be a type-function with respect to $\gamma(t)$ and $\delta(x)$, if $f(x)$ satisfies the following conditions:

1° $f(x)$ is continuous and non-decreasing for $x \geq 0$ and tends to infinity with $x$;

2° $f(x) \geq 0$ for $x \geq 0$;

3° $f(x+\gamma\{f(x)\}) - f(x) < \delta(x)$ for $x \geq 0$.

DEFINITION 11.2. Let $f(x)$ and $g(x)$ be two functions defined for $x \geq 0$. $g(x)$ is said to be superiorly adjoined to $f(x)$, if $f(x) \leq g(x)$ when $x$ is sufficiently large and $f(x) = g(x)$ for a sequence of values of $x$ tending to infinity. $g(x)$ is said to be inferiorly adjoined to $f(x)$, if $f(x) \geq g(x)$ when $x$ is sufficiently large and $f(x) = g(x)$ for a sequence of values of $x$ tending to infinity.

THEOREM 11.2. Let $f(x)$, $\gamma(t)$ and $\delta(x)$ be three functions satisfying the following conditions:

$f(x)$ is continuous and non-decreasing for $x \geq 0$ and tends to infinity with $x$;

$\gamma(t)$ and $\delta(x)$ satisfy the conditions in Theorem 1. Then there exist two functions $f_1(x)$ and $f_2(x)$ having the following properties:

1° $f_1(x)$ and $f_2(x)$ are type-functions with respect to $\gamma(t)$ and $\delta(x)$;

2° $f_1(x)$ and $f_2(x)$ are respectively superiorly and inferiorly adjoined to $f(x)$.

In the proof of Theorem 11.2, the following lemma which itself is a theorem on type-functions, plays an important role:

LEMMA 11.1. Let $f(x)$ be a positive, continuous and non-increasing function for $x \geq 0$ and tending to zero with $1/x$. Then there exist two functions $f_1(x)$ and $f_2(x)$ having the following properties:

1° $f_1(x)$ and $f_2(x)$ are linear by segments, positive and non-increasing for $x \geq 0$ and tend to zero with $1/x$.

2° $f_1'(x) > -1$ and $f_2'(x) > -1$ for $x \geq 0$.

3° $f_1(x)$ and $f_2(x)$ are respectively superiorly and inferiorly adjoined to $f(x)$.

C. An Extension of Theorem 11.1

In Theorem 11.1 the function $f(x)$ is supposed to be continuous for $x \geq 0$. In a subsequent work [90] I got rid of this conditions and proved the following theorem:

THEOREM 11.1'. Let $f(x)$, $\gamma(t)$ and $\delta(x)$ be three functions satisfying the following conditions:

$f(x)$ is non-decreasing and $f(x) \geq t_0$ for $x \geq x_0$.

$\gamma(t)$ is positive, continuous and non-increasing for $t \geq t_0$, $\int_{t_0}^{\infty} \gamma(t)dt$ converges.

$\delta(x)$ is positive, continuous and non-increasing for $x \geq x_0$, $\int_{x_0}^{\infty} \gamma(x)dx$ diverges.

Then the exterior measure of the set

$$\{x \mid x \geq x_0,\ f(x+\gamma\{f(x)\}) - f(x) \geq \delta(x)\}$$

with respect to the function $\int_{x_0}^{x} \delta(x)dx$ does not exceed $\delta(x_0)\gamma(t_0) + \int_{t_0}^{\infty} \gamma(t)dt$.

In general, $U(x)$ being a continuous increasing function in an interval $x \geq x_0$ such that

$$\lim_{x\to+\infty} U(x) = +\infty$$

and $E$ a set of points in the interval $x \geq x_0$, the exterior measure of $E$ with respect to the function $U(x)$ is, by definition, the exterior measure

$$m^*G = \lim_{Y\to+\infty} m^*(G \cap [y_0 Y]),\ y_0 = U(x_0)$$

of the set $G$ of the values which the function $U(x)$ takes on $E$.

We are going to give some applications of Theorem 11.1'.

First application.

First of all, in taking the logarithm of both sides of the inequality in Borel's theorem, we see that the following theorem should be true:

THEOREM 3. Let $f(x)$ be a positive and non-decreasing function for $x \geq x_0$ such that

$$\lim_{x\to+\infty} f(x) = +\infty.$$

Let $\lambda$ and $\mu$ be two positive numbers. Then the exterior measure of the set

$$\{x | x \geq x_0, \ f(x + \frac{1}{f(x)^\lambda}) \geq (1+\mu)f(x)\}$$

is finite.

In fact, for the proof of this theorem, it is sufficient to apply Theorem 1' to the functions $\log\{f(x)\}(x \geq x_1), \gamma(t) = 1/t^2 (t \geq 1)$ and $\delta(x) = \log(1+\mu)$, where $x_1 \geq x_0$ is such that

$$1 \leq \log\{f(x)\} \leq f(x)^{\lambda/2}$$

for $x \geq x_1$.

In 1980, when Hayman visited Peking, he said, in a conversation, that in his research on the distribution of the values of meromorphic functions, he needs the following theorem:

THEOREM 11.4. Let $f(x)$ be a positive and non-decreasing function for $x \geq x_0$ such that

$$\lim_{x\to+\infty} f(x) = +\infty.$$

Let $\alpha$ and $\beta$ be two numbers such that $\alpha > 0$, $\alpha + \beta > 1$. Then the exterior measure of the set

$$\{x | x \geq x_0, \ f(x + \frac{1}{f(x)^\alpha}) \geq f(x) + f(x)^\beta\}$$

is finite.

If we write the inequality in Theorem 11.3 in the form

$$f(x + \frac{1}{f(x)^\lambda}) \geq f(x) + \mu f(x),$$

we see that Theorem 11.4 is a generalization of Theorem 3. Based upon Theorem 11.1' and Theorem 11.3, Chuang gave a proof of Theorem 11.4 in distinguishing four cases: $\beta \geq 1$, $\beta = 0$, $\beta < 0$, $0 < \beta < 1$ [91].

Chuang's work was followed by that of Wang [92]. He proved by an entirely different method the following theorem:

THEOREM 11.4'. Let $\alpha$ be a positive number and $g(u)$ a positive function for $u \geq u_0$ $(u_0 > 0)$ such that the function $u^\alpha g(u)$ is non-decreasing for $u \geq u_0$ and $\int_{u_0}^{\infty} \frac{du}{u^\alpha g(u)}$ converges. Let $f(x) \geq u_0$ be a non-decreasing function for $x \geq x_0$ such that

$$\lim_{x \to +\infty} f(x) = +\infty.$$

Then the exterior measure of the set

$$\{x \mid x \geq x_0,\ f(x + \frac{1}{f(x)^\alpha}) \geq f(x) + g\{f(x)\}\}$$

is finite.

In the proof, he used the following covering theorem [93]:

Let $E \subset (-\infty,\infty)$ be a set of points and $\{B_v\}$ $(v \in V)$ a family of closed intervals. Suppose that the length $mB_v$ of each interval $B_v$ does not exceed a fixed constant $K$ and that $E \subset \bigcup_{v \in V} B_v$. Then there exists a sequence (finite or infinite) of mutually disjoint intervals $B_1, B_2, \ldots, B_k, \ldots$ of the family $\{B_v\}$ such that

$$\sum_{k=1}^{\infty} mB_k \geq \frac{1}{5} m^*E.$$

More generally, we may consider two positive functions $a(u)$ and $b(u)$ $(u \geq u_0)$ and try to find conditions such that if the functions $a(u)$ and $b(u)$ satisfy these conditions, then for every function $f(x)$ which is non-decreasing and $f(x) \geq u_0$ for $x \geq x_0$ and is such that

$$\lim_{x \to +\infty} f(x) = +\infty,$$

the exterior measure of the set

$$\{x \mid x \geq x_0,\ f(x + \frac{1}{a\{f(x)\}}) \geq f(x) + b\{f(x)\}\}$$

is finite.

It can be shown that if the positive functions $a(u)$ and $b(u)$ are continuous and such that the function $a(u)b(u)$ is non-decreasing, then the condition

$$\int_{u_0}^{\infty} \frac{du}{a(u)b(u)} < +\infty$$

is necessary, but not sufficient.

An interesting problem is to find a necessary and sufficient condition.

Second application.

Consider a non-constant entire function

$$f(z) = \sum_{n=0}^{\infty} a_n z^n$$

and set, for $r > 0$,

$$M(r) = \max_{|z|=r} |f(z)|, \quad A(r) = \max_{|z|=r} \operatorname{Re}\{f(z)\}$$

$$m(r) = \max_{n\geq 0} |a_n| r^n, \quad T(r) = \frac{1}{2\pi}\int_0^{2\pi} \log^+|f(re^{i\theta})|\,d\theta,$$

$$M'(r) = \max_{|z|=r} |f'(z)|.$$

We know that the following inequalities hold for $0 < r < R$: (see [1], [94], (87]).

$$T(r) \leq \log^+ M(r) \leq \frac{R+r}{R-r} T(R),$$

$$A(r) \leq M(r) \leq |f(0)| + \frac{2r}{R-r} A(R) - \operatorname{Re}\{f(0)\}\},$$

$$m(r) \leq M(r) \leq \frac{R}{R-r} m(R),$$

$$M(r) - |f(0)| \leq rM'(r) \leq \frac{r}{R-r} M(R).$$

On the basis of these inequalities and Theorem 11.1', various results can be obtained on the asymptotic behavior of the functions

$$\frac{\log M(r)}{T(r)}, \quad \frac{M(r)}{A(r)}, \quad \frac{M(r)}{m(r)}, \quad \frac{rM'(r)}{M(r)}$$

as $r \to +\infty$ through certain sets of values of the interval $r > 0$ [95]. Here we mention only one of these results, namely

$$\lim_{r\to+\infty} \frac{\log M(r)}{\tau(r)} = 0, \quad \tau(r) = \frac{T(r)}{\gamma\{\log T(r)\}},$$

where the function $\gamma(t)$ satisfies the conditions in Theorem 11.1'. In particular, taking

$$\gamma(t) = \frac{1}{t^\lambda} \quad (\lambda > 1,\ t \geq 1),$$

we get the known relation [13]:

$$\lim_{r\to+\infty} \frac{\log M(r)}{T(r)\{\log T(r)\}^{\lambda}} = 0$$

D. Extensions of Theorem 11.2 and Lemma 11.1

In Theorem 11.2 the function $f(x)$ is supposed to be non-decreasing and tending to infinity with $x$. For certain applications, it is convenient to replace this condition by the weaker condition.

$$\overline{\lim_{x\to+\infty}}\ f(x) = +\infty.$$

We have the following theorem [96] [97]:

THEOREM 11.2'. Let $f(z)$, $\gamma(t)$ and $\delta(x)$ be three functions satisfying the following conditions:

$f(x)$ is continuous for $x \geq x_0$ and $\overline{\lim}_{x\to+\infty} f(x) = +\infty$.

$\gamma(t)$ and $\delta(x)$ satisfy the conditions in Theorem 1'.

Then there exists a function $F(x)$ having the following properties:

1° $F(x)$ is a type-function with respect to $\gamma(t)$ and $\delta(x)$.[3)]

2° $F(x)$ is superiorly adjoined to $f(x)$.

This theorem is useful for the construction of certain type-functions. For example, consider a meromorphic function $\phi(z)$ of infinite order. Then, by definition, we have

$$\overline{\lim_{r\to+\infty}} \frac{\log T(r,\phi)}{\log r} = +\infty.$$

In order to show that the function $\phi(z)$ has an order $\rho(r)$ in the sense of Hiang mentioned above, consider the function

$$\tau(r) = \frac{\log T(r,\phi)}{\log r} \qquad (r \geq r_0)$$

and apply Theorem 11.2' to the three functions

$$f(x) = \log \tau(e^x)(x \geq x_0, x_0 > 0),\ \gamma(t) = \frac{1}{t^2}\ (t \geq 1),\ \delta(x) = \frac{1}{x}\quad (x \geq x_0).$$

Then it is easy to verify that the function

$$\rho(r) = e^{F(\log r)} \quad (r \geq r_0)$$

---

3) According to a slightly modified form of Definition 1.1.

satisfies the required conditions, where $F(x)$ is the function furnished by Theorem 11.2'.

Similarly we have the following extension of Lemma 11.1 [96],[97]:

LEMMA 11.1'. Let $f(x)$ be a continuous function for $x \geq x_0$ such that $\overline{\lim\limits_{x\to+\infty}} f(x) = 0$. Let $\eta(x)$ be a positive continuous function for $x \geq x_0$ such that $\int_{x_0}^{\infty} \eta(x)dx$ diverges. Then there exists a function $F(x)$ satisfying the following conditions:

1° $F(x)$ is monotonous and continuously differentiable by sections for $x \geq x_0$ with $\lim\limits_{x\to+\infty} f(x) = 0$.

2° $|F'(x)| < \eta(x) \quad (x \geq x_0)$.

3° $F(x)$ is superiorly adjoined to $f(x)$.

From this lemma we deduce immediately the existence of a precise order $\rho(r)$ in the sense of Valiron of a meromorphic function $\phi(z)$ of finite positive order $\rho$. In fact, since, by definition, we have

$$\overline{\lim_{r\to+\infty}} \frac{\log T(r,f)}{\log r} = \rho,$$

we can apply Lemma 11.1' to the functions

$$f(r) = \frac{\log T(r,f)}{\log r} - \rho \quad (r \geq r_0), \quad \eta(r) = \frac{1}{r(\log r)(\log \log r)} \quad (r \geq r_0)$$

and we verify easily that the function

$$\rho(r) = \rho + F(r) \quad (r \geq r_0)$$

is a precise order of the function $\phi(z)$, where $F(r)$ is the functional furnished by Lemma 11.1'.

### E. Pólya Peaks

The notion of Pólya peaks originates from the following theorem of Pólya [98] on sequences of positive numbers:

Let $\ell_m (m = 1,2,\ldots)$ be a sequence of positive numbers and $S_m (m = 1,2,\ldots)$ an increasing sequence of positive numbers. If

$$\lim_{m\to+\infty} \ell_m = 0, \quad \overline{\lim_{m\to+\infty}} \ell_m S_m = +\infty,$$

then we can find arbitrarily large integers $n$ such that

$$\ell_n > \ell_\nu \; (\nu > n), \quad \ell_n S_n > \ell_\mu S_\mu \quad (\mu < n).$$

The corresponding theorem on functions of a continuous variable was first obtained by Shah [99], [100], and then independently by Edrei and Fuchs [101]. Afterwards, Edrei gave the definition of a sequence of Pólya peaks of a given order and proved its existence [102], [103]. In modern theory of meromorphic functions, the notion of Pólya peaks plays an important role. In what follows, our main purpose is to point out that the notion of Pólya peaks is closely related to that of type-functions. In fact, we shall see that the existence of certain sequences of Pólya peaks can easily be deduced from Lemma 11.1' which is a theorem on type-functions [97].

We first give the following definition of sequences of Pólya peaks, which differs in form from that given by Edrei.

DEFINITION 11.3. Let $f(x)$ and $U(x)$ be two functions defined in an interval $x \geq a$, in which $U(x)$ is increasing and unbounded. We say that a sequence $x_n (n = 1,2,\ldots)$ is a sequence of upper Pólya peaks of $f(x)$ with respect to $U(x)$, if

$$x_n > a (n = 1,2,\ldots), \quad \lim_{n\to+\infty} x_n = +\infty,$$

and there exist three sequences $x_n'$, $x_n''$, $\varepsilon_n$ $(n = 1,2,\ldots)$ such that

1° $a < x_n' < x_n < x_n''$, $\varepsilon_n \geq 0$ $(n = 1,2,\ldots)$,

2° as $n \to +\infty$, $x_n'$, $U(x_n) - U(x_n')$ and $U(x_n'') - U(x_n)$ tend to $+\infty$ and $\varepsilon_n \to 0$,

3° for each $n$, we have

$$f(x) \leq f(x_n) + \varepsilon_n \quad (x_n' < x < x_n''). \tag{11.1}$$

Similarly, we can give the definition of a sequence of lower Pólya peaks of $f(x)$ with respect to $U(x)$; the only difference is to replace (11.1) by

$$f(x) \geq f(x_n) - \varepsilon_n \quad (x_n' < x < x_n'') \tag{11.2}$$

In particular, if in (11.1) we have $\varepsilon_n = 0$, then in the interval $x_n' < x < x_n''$ the graph of $f(x)$ looks like

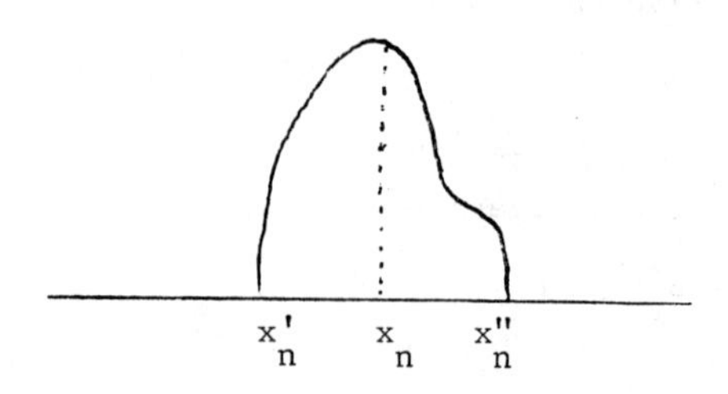

Similarly, if in (11.2) we have $\varepsilon_n = 0$, then the graph looks like

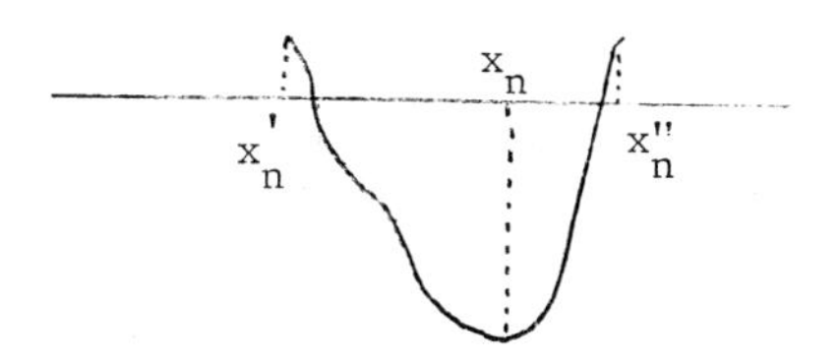


Of course, if $\varepsilon_n > 0$, the form of the graph might be quite different.

On the other hand, it is easy to verify that if in Definition 11.3, the function $f(x)$ is of the particular form $f(x) = \log \phi(x) - \rho \log x$, where $\phi(x)$ is a positive function and $\rho \geq 0$ a number, then a sequence of upper Pólya peaks of $f(x)$ is a sequence of Pólya peaks of order $\rho$ of $\phi(x)$ according to the definition given by Edrei. The converse is also true.

Now consider a continuous function $g(x)$ for $x \geq x_0$ $(x_0 > 1)$ such that

$$\overline{\lim_{x\to+\infty}} \frac{g(x)}{x} = 0$$

Then we can apply Lemma 1' to the functions

$$f(x) = \frac{g(x)}{x}, \quad \eta(x) = \frac{1}{x \log x} \qquad (x \geq x_0).$$

Hence there exists a function $F(x)$ satisfying the conditions in Lemma 11.1'. By the third condition, there exist a value $X > x_0$ and a sequence $x_n$ $(x_n > 2X)$ $(n = 1,2,\ldots)$ tending to $+\infty$ such that

$$g(x) \quad xF(x) \qquad (x \geq X),$$

$$g(x_n) = x_n F(x_n) \qquad (n = 1,2,\ldots).$$

By simple calculations, it is found that the sequence $x_n$ $(n=1,2,\ldots)$ is a sequence of upper Pólya peaks of $g(x)$ with respect to $x$.

From this result we deduce easily the following theorem.

THEOREM 11.5. Let $f(x)$ be a positive, continuous function for $x \geq x_0$ $(x_0 > 1)$. Then the following assertions hold:

1° If $\quad = \overline{\lim_{x\to+\infty}} \dfrac{\log f(x)}{\log x}$ is finite, the function $\log f(x) - \rho \log x$ has a sequence of upper Pólya peaks with respect to $\log x$.

2° If $\rho' = \underline{\lim_{x\to+\infty}} \dfrac{\log f(x)}{\log x}$ is finite, the function $\log f(x) - \rho' \log x$ has a sequence of lower Pólya peaks with respect to $\log x$.

REFERENCES

1. Nevanlinna, R., Le théoreme de Picard-Borel et la théorie des fonctions meromorphes, Paris, 1929.

2. Dufresnoy, J., Sur les valeurs exceptionnelles des fonctions méromorphes voisines d'une fonction meromorphe donnee, C. R. Acad. Sc., 208 (1939), 255.

3. Chuang, Chi-tai, Une generalisation d'une inegalite de Nevanlinna, Scientia Sinica, 13 (1964), 887-895.

4. _______, On the distribution of the values of meromorphic functions (in Chinese), Chinese Annals of Mathematics, 1 (1980), 93-114.

5. _______, On the distribution of the values of meromorphic functions, Kexue Tongbao, 25 (1980), 625-630.

6. Yang, Lo , Deficient functions of meromorphic functions, Sci. Sinica, 24 (1981), 1179-1189.

7. Baernstein, A., Proof of Edrei's spread conjecture, Proc. London Math. Soc., 26 (1973), 418-434.

8. Edrei, A., Solution of the Deficiency problem for functions of small lower order, Proc. London Math. Soc., 26 (1973), 435-445.

9. Chen, Ding-xing, On Valiron deficient functions, Scientia Sinica, 10 (1982), 878-889.

10. Hyllengren, A., Valiron deficient values for mermorphic functions in the plane, Acta Math., 124 (1970), 1-8.

11. Rauch, A., Extensions de theorèmes relatifs aux directions de Borel des functions meromorphes, Journ. de Math., 12 (1934).

12. Pfluger, A., Zur Defektrelation ganzer Functionen endlicher ordnung, Comment. Math. Helv., 19 (1946), 91-104.

13. Edrei, A. and Fuchs, W.H.J., Valeurs deficientes et valeurs asymptotiques des fonctions meromorphes, Comment. Math. Helv., 33 (1959), 317-344.

14. Hayman, W.K., Meromorphic Functions, Oxford, 1964.

15. Yang, Lo and Zhang, Guanghou, Deficient values of extremal entire functions, Sci. Sinica, 20 (1977), 421-435.

16. _______, Deficient values and asumptotic values of entire functions, Sci. Sinica, Special issue (II) (1979), 190-203.

17. _______, Deficient values of a meromorphic function and all its derivatives, Acta Math. Sinica, 25 (1982), 617-625.

17a. _______, Distribution of zeros and poles of meromorphic functions and their filling disks, Sci. Sinica, Series A, 25 (1982), 371-383.

18. Goldberg, A. A., On the deficiencies of meromorphic functions, Dokl. Akad. Nauk SSSR, 98 (1954), 893-895.

19. Arakelyan N. U., Entire functions of finite order with an infinite set of deficient values, Dokl. Adad. Nauk SSSR, 170 (1966), 999-1002.

20. Drasin, D., Zhang, Guanghou, Yang, Lo and Weitsman, A., Deficient values of entire functions and their derivatives, Proc. Amer. Math. Soc., 82 (1981), 607-612.

21. Miller, S., Complex analysis (proc. Conf. S.U.N.Y. Brockport, 1978), Dekker, New York, 1978.

22. Valiron, G., Directions de Borel des Fonctions méromorphes, Mem. Sci. Math., fasc. 89, Paris, 1938.

23. Yang, Lo et Zhang, Guanghou, Recherches sur le nombre des valeurs deficientes et le nombre des directions de Borel des fonctions méromorphes, Sci. Sinica, 18 (1975), 23-37.

24. _______, Distribution of Borel directions of entire functions, Acta Math. Sinica (in Chinese), 19 (1976), 157-168.

25. _______, A general theorem on total numbers of deficient values of entire functions, Acta. Math. Sinica (in Chinese), 25 (1982), 244-247.

26. Zhang, Guanghou, On relations between deficient values, asymptotic values and Julia directions of entire and meromorphic functions, Sci. Sinica, Supplement 1, (1978).

27. Yang, Lo et Zhang, Guanghou, Sur la construction des fonctions meromorphes ayant des direction singulieres donnees, Sci. Sinica, 19 (1976), 445-459.

28. _______, Sur la distribution des directions de Borel des fonctions méromorphes, Sci. Sinica, 16 (1973), 465-482.

29. _______, Distribution of Borel directions of entire functions, Acta Math. Sinica (Chinese), 19 (1976), 157-168.

30. Rauch, A., Cas ou une direction de Borel d'une fonction entiere f(z) est aussi direction de Borel pour f'(z), C.R. Acad. Sci., 199 (1934), 1014.

31. Chuang, Chi-tai, Un théorème relatif aux directions de Borel des fonctions meromorphes d'ordre fini, C.R. Acad. Sci., 204 (1937), 951.

32. _______, Sur les suites de cercles de remplissage et les directions de Borel des fonctions méromorphes, Sci. Sinica, 22 (1979), 493-511.

33. Zhang, Guanghou, Research on common Borel directions of a meromorphic function and of its successive derivatives and integrals (I), (II), (III) (in Chinese). Acta Math. Sinica, 20 (1977), 73-98, 157-177, 237-247.

34. Milloux, H., Sur les directions de Bo-el des fonctions entieres, de leurs derivees et de leurs integrales, J. d'analyse math., 1 (1951), 244-330.

35. Yang, Lo, Common Borel directions of meromorphic functions and their derivatives, Sci. Sinica, Special Issue (II), (1979), 91-104.

36. Yang, Lo and Zhang Qingde, New singular direction of meromorphic functions, Sci. Sinica (Series A), 27 (1984), 352-366.

37. Bureau, F., Memoire sur les fonctions uniformes a point singulier essentiel isole. Mémoires de la Societe Royale des Sciences de Liége, 17 (1932).

38. Miranda, C., Sur un nouveau critere de normalite pour les familles des fonctions holomorphes, Bull. de la Societe Mathématique, 63 (1935) 185-196.

39. Valiron, G., Sur les valeurs exceptionnelles des fonctions, meromorphes et de leurs derivees, Actualites Scientifiques et Industrielles, 570 (1937).

40. Chuang, Chi-tai, Ètude sur les familles normales et les familles quasi-normales de fonctions mèromorphès, Rendiconti circolo Mat. Palermo, 62 (1938).

41. _______, Un théorème genéral sur les fonctions homomorphes dans le cercle unite et ses applications (I), (II), Scientia Sinica, 6 (1957), 569-621, 757-831.

42. Milloux, H., Les fonctions mermorphes et leurs derivées, Actualites Scientifiques et Industrielles, 888 (1940).

43. _______, Sur une nouvelle extension d'une inégalité de M. R. Nevanlinna, J. Math. pures et appl., 19 (1940), 197-210.

44. Hayman, W.K., Picard values of meromorphic functions and their derivatives, Ann. of Math., 70 (1959), 9-42.

45. Ku Yunghsing, Un critère de normalité des familles de fonctions méromorphes, Sci. Sinica, Special Issue (I), (1979), 267-274.

46. Yang, Lo, Meromorphic functions and their derivatives, J. London Math. Soc., 25 (1982), 288-296.

47. _______, A fundamental inequality and its application, Chin. Ann. of Math. (Series B), 4 (1983), 339-346.

48. Yang, Lo et Zhang, Guanghou, Recherches sur la normalité des familles de fonctions analytiques à des valeurs multiples Sci. Sinica, 14 (1965), 1258-1271.

49. _______, Recherches sur la normalité des familles de fonctions analytiques a des valeurs multiples II, Sci. Sinica, 15 (1966), 433-453.

49a. Oshkin, I. B., On a condition for the normality of families of holomorphic functions, Uspekhi. Mat. Nauk, 37 (1982), 221-222.

50. Ku, Yunghsing, Sur les familles normal des fonctions meromorphes, Sci. Sinica, 4 (1978).

51. Yang, Lo, Normal families and differential polynomials, Sci. Sinica (Series A), 26 (1983), 673-686.

51a. Li Xianjin, Proof of a conjecture of Hayman, Sci. Sinica (to appear).

52. Iversen F., Recherches sur les fonctions inverses de fonctions méromorphes, Thèse de Helsingtors, 1914.

53. Ahlfors, L. V., Untersuchungen zur Theorie der konformen Abbildung und der ganzen Funktionen, Acta Soc. Sci. Fenn. Nova Ser. I, no. 9 (1930).

54. Hayman, W.K., Research Problems in Function Theory, 1967, London, p. 11.

55. Huber, A., On subharmonic functions and differential geometry in the large, Comment. Math. Helv., 32 (1957), 13-72.

56. Zhang, Guanghou, Asymptotic values of entire and meromoprhic functions, Sci. Sinica, 20 (1977), 720-739.

57. ______, On the length of asymptotic path of an entire function (in Chinese), Sci. Sinica, 1979, no. 7, 641-648.

58. ______, On entire functions extremal for Denjoy conjecture (in Chinese), Sci. Sinica, 1980, no. 12, 1125-1136.

59. ______, On entire functions extremal for Denjoy conjecture (in Chinese), Sci. Sinica, 1982, no.11, 983-994.

60. Valiron, G., Points de Picard et Points de Borel des functions meromorphes dans un cercle, Bull. Sci. Math., 2 e serie, 56 (1932), 10-32.

61. Yang, Lo et Shiao, Shiou-zhi, Sur les points de Borel des fonctions méromorphes et de leurs derivees, Sci. Sinica, 14 (1965), 1556-1573.

62. Xing, Hua-lin, On filling circles of meromorphic functions in the unit disk (in Chinese), Chinnese Annales of Mathematics, 2 (1981), 497-510.

63. Hayman, W.K. and Yang, Lo, Growth and values of functions regular in an angle, Proc. London Math. Soc., 44 (1982), 193-214.

64. Lai, Wan-zai, Precise value of Hayman, constant in Landau theorem (in Chinese), Sci. Sinica, 1978, no. 5, 495-500.

65. Zhao, Jingyi, On entire algebroid functions with two branches, Mat. cb., 37 (79) (1955), 573-576.

66. Hiong, King-lai, Sur la croissance des fonctions algébroides en rapport avec leurs dérivees, C.R. Acad. Sc., 241 (1956), 3032-3035.

67. ______, Sur les fonctions méromorphes et les fonctions algébroides, Mem. Sci. Math., Fasc. 139, Paris (1957).

68. He, Yuzan, Sur un problème d'unicite relatif aux fonctions algebroides, Sci. Sinica, 14 (1965), 174-180.

69. ______, Sur les fonctions algebroides et leurs dérivees, Acta Math. Sinica, 15 (1965), 500-510.

70. ______, On the mulriple values of algebroid functions, Acta Math. Sinica, 22 (1979), 733-742.

71. Lu, Yirian, On direct transcendental singularities of the inverse function of an algebroidal function, Sci. Sinica, 23 (1980), 407-415.

72. Rauch, A., Sur les algebroides entières, C.R. Acad. Sci., Paris 202 (1936), 2041-2042.

73. Toda, N., Sur les directions de Julia et de Borel des fonctions algebroides Nagoya Math. J. 34 (1969), 1-23.

74. Lu, Yinian, On Borel direction of an algebroidal function, Sci. Sinica, Series A 25 (1982), 25-30.

75. Lu, Yinian and Gu Yongxing, On the existence of Borel direction of an algebroidal function Kexue Tongbao, 28 (1983), 264-266.

76. Ozawa, M., On the growth of algebroid functions with several deficiencies, Kodai Math. Sem. Rep. 22 (1970), 122-127.

77. Gu, Yongxing, The growth of algebroid functions with several deficient values, Contemporary Math. 25 (1983), 45-49.

78. Gackstatter, F. and Laine, I., Zur Theorie der gewohnlichen Differentialgleichungen in Komplexen, Ann. Pol. Math., 38 (1980), 259-287.

79. He, Yuzan, On algebroid solutions of ordinary differential equations, Acta Math., Sinica, 24 (1981), 464-471.

80. He, Yuzan and Xiao Xiuzhi, On Malmquist's theorem of algebraic differential equations, Kexue Tongbao, 28 (1983), 165-169.

81. _______, Meromorphic and algebroid solution of higher order algebraic differential equations, Sci. Sinica, series A. 26 (1983), 1034-1043.

82. _______, Admissible solutions of ordinary differential equations, Contemporary Math. 25 (1983), 51-61.

83. Yosida, K., On algebroid-solutions of ordinary differential equations, Japan J. Math., 10 (1934), 199-208.

84. Borel, E., Sur les zeros des fonctions entières, Acta Math., 20 (1897).

85. Blumenthal, O., Principes de la théorie des fonctions entières d'ordre infini, Paris, Gauthier-Villars, (1910).

86. Hiong, King-Lai, Sur les fonctions entières et les fonctions méromorphes d'ordre infini, J. de Math., 14, (1935).

87. Valiron, G., Lectures on the general theory of integral functions, Toulouse, Edouard Privat, 1923.

88. Nevanlinna, R., Remarques sur les fonctions monotones, Bull. Sci. Math., 55 (1931).

89. Chuang, Chi-tai, Sur les fonctions continues monotones, Ann. Ecole Norm., 64 (1947).

90. _______, On monotonous functions (in Chinese) J. Peking Univ. 1 (1955).

91. _______, A generalization of a theorem of Borel on monotone functions, Kexue Tongbao, 27 (1982).

92. Wang, Shi-ra, Two theorems on monotonous functions, Kexue Tongbao, 1983.

93. Stein, E.M. Singular integrals and differentiability properties of functions, Princeton, (1970).

94. Titchmarsh, E.C., The theory of functions, Oxford University press, 1939.

95. Chuang, Chi-tai, Sur la croissance de fonctions, Kexue Tongbao, 26 (1981).

96. _______, Sur les fonctions-types, Sci. Sinica, 10 (1961).

97. _______, On Polya peaks, Sci. Sinica, 23 (1980).

98. Polya, G., and Szego, G., Aufgaben und Lehrsatze aus der Analysis, Berlin, 1964.

99. Shah, S.M., A theorem on integral functions of integral order, J. London Math. Soc., 15 (1940).

100. _______, A theorem on integral functions of integral order, II, J. Indian Math. Soc., N.S., 5 (1941).

101. Edrei, A. and Fuchs, W.H.J., The deficiencies of meromorphic functions of order less than one, Duke Math. J., 27 (1960).

102. Edrei, A., Sums of deficiencies of meromorphic functions, J. Analyse Math., 14 (1965), 79-107.

103. _______, Locally Tauberian Theorems for meromorphic functions of lower order less than one, Trans. Amer. Math. Soc., 140 (1969).

Department of Mathematics
Peking University
Beijing

Institute of Mathematics
Academia Sinica
Beijing

Contemporary Mathematics
Volume 48, 1985

# A SIMPLE PROOF OF A THEOREM OF FATOU ON THE ITERATION AND FIX-POINTS OF TRANSCENDENTAL ENTIRE FUNCTIONS

Chi-tai Chuang

In the theory of Fatou [1] on the iteration and fix-points of transcendental entire functions, the following theorem is fundamental:

THEOREM. Let $f(z)$ be a transcendental entire function and let the sequence of iterates $f_n(z)(n = 1,2,\dots)$ be defined by

$$f_1(z) = f(z),\ f_{n+1}(z) = f_1\{f_n(z)\}(n =1,2,\dots). \tag{1}$$

Denote by $E(f)$ the set of those points of the complex plane where the family $\{f_n(z)\ (n = 1,2,\dots)\}$ is not normal and by $H(f)$ the set of fix-points of all orders of $f(z)$. Then the following assertions hold:

$1^0$ $E(f)$ is a non-empty perfect set.

$2^0$ $E(f)$ belongs to the derived set of $H(f)$.

The purpose of this paper is to give a proof of this theorem by a method which, apart from several well known theorems of Montel on normal families [2], bases mainly on Nevanlinna's theory of meromorphic functions [3]. We shall use the symbols:

$$T(r,f) = \frac{1}{2\pi}\int_0^{2\pi}\log^+|f(re^{i\theta})|d\theta,\ M(r,f) = \max_{|z|=r}|f(z)|$$

for an entire function $f(z)$.

For the proof we need the following lemmas:

LEMMA 1. There exists an absolute constant $A > 0$ such that if $f(z)$ and $g(z)$ are transcendental entire functions and if $r > 0$ is a number satisfying the inequality

$$AM(\tfrac{r}{4},f) > |f(0)|(A + 1),$$

then we have

$$T(r,\phi) > \frac{1}{3}T(\rho,g),$$

where

$$\phi(z) = g\{f(z)\},\ \rho = AM(\tfrac{r}{4},f) - |f(0)|(A + 1).$$

The proof of this lemma can be found in a book of Hayman [4].

LEMMA 2. Let $f(z)$ be a transcendental entire function and $f_n(z)$ $(n = 1,2,\ldots)$ be the sequence of iterates defined by (1). Let $K > 1$ be a number. Then we can find a number $r_0 = r_0(f,K) > 1$ such that for $n \geq 1$ and $r > r_0$, we have

$$T(r,f_n) > \frac{1}{2}\left(\frac{K}{3}\right)^{n-1} T(r,f). \tag{2}$$

This lemma due to Zhang Guanghou is unpublished. It plays an important role in the proof of the above theorem.

Proof: $A$ being the constant occuring in Lemma 1, let $r_1 > 1$ be a number such that for $r > r_1$, we have

$$AM\left(\frac{r}{4},f\right) - |f(0)|(A+1) > r^K.$$

Then by applying Lemma 1 to the functions $f(z)$ and $g(z) = f_{n-1}(z)$, we see that for $n > 1$ and $r > r_1$, we have

$$T(r,f_n) > \frac{1}{3}T(r^K,f_{n-1}).$$

Since $r^K > r$, we have also

$$T(r^K,f_{n-1}) > \frac{1}{3}T(r^{K^2},f_{n-2}),$$

and so on. Finally we get the inequality

$$T(r,f_n) > \left(\frac{1}{3}\right)^{n-1} T(r^{K^{n-1}},f). \tag{3}$$

Next by the convexity of the function $T(r,f)$ with respect to $\log r$, the function

$$\frac{T(r,f) - T(1,f)}{\log r}$$

in non-decreasing for $r > 1$. Consequently for $n > 1$ and $r > 1$, we have

$$\frac{T(r^{K^{n-1}},f) - T(1,f)}{K^{n-1}\log r} \geq \frac{T(r,f) - T(1,f)}{\log r}. \tag{4}$$

Let $r_2 > 1$ be a number such that for $r > r_2$, we have

$$T(r,f) > 2T(1,f).$$

Then from (3) and (4), we see that the number $r_0 = \max(r_1,r_2)$ has the property required in Lemma 2.

Now let us return to the proof of the theorem. We first show that the set $E(f)$ contains an infinite number of points. Consider the number $r_0 = r_0(f,4)$ for $K = 4$ occuring in Lemma 2 and a number $R \geq r_0$. Evidently it is sufficient to show that the family $\{f_n(z)(n = 1,2,\ldots)\}$ is not normal in the domain $D: R < |z| < +\infty$. In fact, by a theorem of Rosenbloom [5], the function $f_2(z)$ has an infinite number of fix-points of which there must be one $\zeta \in D$. Then

$$f_{2p}(\zeta) = \zeta,\ f_{2p+1}(\zeta) = f(\zeta)\ (p = 1,2,\ldots)$$

and the sequence $f_n(\zeta)$ $(n = 1,2,\ldots)$ is bounded. Consequently, by a well known theorem on normal families [2], if the family $\{f_n(z)\ (n = 1,2,\ldots)\}$ is normal in $D$, then the sequence $f_n(z)$ $(n = 1,2,\ldots)$ is locally uniformly bounded in $D$. Consider a circle $|z| = r$ with $r > R$. There is then a number $B > 0$ such that

$$M(r,f_n) \leq B \quad (n = 1,2,\ldots). \tag{5}$$

On the other hand, by Lemma 2, we have

$$T(r,f_n) > \frac{1}{2}\left(\frac{4}{3}\right)^{n-1} T(r,f) \quad (n = 1,2,\ldots) \tag{6}$$

(5) and (6) being incompatible, we arrive at a contradiction.

Let $z_0$ be a point of $E(f)$. To see that $z_0$ is a point of accumulation of $E(f)$, consider a circle $C: |z-z_0| < r$. Since the family $\{f_n(z)$ $n = 1,2,\ldots)\}$ is not normal in $C$, there is a sequence $f_{n_k}(z)$ $(k = 1,2,\ldots)$ from which we cannot extract a subsequence converging locally uniformly in $C$ to a holomorphic function or the constant $\infty$. Consider the sequence $f_{n_k}(z_0)$ $(k = 1,2,\ldots)$ and distinguish two cases:

$1^0$ The sequence $f_{n_k}(z_0)$ $(k = 1,2,\ldots)$ is bounded. Then we can find two points $\alpha$ and $\beta$ $(\alpha \neq \beta)$ of $E(f)$ such that

$$|f_{n_k}(z_0)| < \min(|\alpha|,|\beta|)\ (k = 1,2,\ldots).$$

Since the family $\{f_{n_k}(z)\ (k = 1,2,\ldots)\}$ is not normal in $C$, there is an integer $k$ such that the function $f_{n_k}(z)$ takes at least once of the values $\alpha$ and $\beta$, say $\alpha$, in $C$, by a well known theorem of Montel [2]. Acoordingly there is a point $z'$ such that

$$z' \in C,\ z' \neq z_0,\ f_{n_k}(z') = \alpha.$$

Making use of the fact that $E(f) = E(f_{n_k})$ and that $E(f_{n_k})$ is completely invariant, we conclude that $z' \in E(f)$.

$2^0$ The sequence $f_{n_k}(z_0)$ $(k = 1,2,\ldots)$ is unbounded. Let $\alpha_1$ and $\beta_1$ $(\alpha_1 \neq \beta_1)$ be two points of $E(f)$. Then we can find a subsequence $f_{m_h}(z_0)$ $(h = 1,2,\ldots)$ of the sequence $f_{n_k}(z_0)$ $(k = 1,2,\ldots)$, such that

$$|f_{m_h}(z_0)| > \max(|\alpha_1|,|\beta_1|) \quad (h = 1,2,\ldots).$$

Since the family $\{f_{m_h}(z)\ (h = 1,2,\ldots)\}$ is also not normal in $C$, we see as above that there is a point $z'$ such that $z' \in C$, $z' \neq z_0$, $z' \in E(f)$.

Conversely it is evident that every point of accumulation of $E(f)$ is a point of $E(f)$. Hence $E(f)$ is a perfect set.

Let $z_0$ be a point of $E(f)$. We are going to show that $z_0$ is a point of accumulation of the set $H(f)$. Consider a circle $C:|z-z_0| < r$. Since $z_0$ is a point of accumulation of $E(f)$, we can find four distinct points $a_j$ $(j = 1,2,3,4)$ in $C$ such that

$$a_j \neq z_0,\ a_j \in E(f) \quad (j = 1,2,3,4).$$

By a theorem of Nevanlinna [3], the function $f(z)$ has at most two completely multiple values. Consequently among the four values $a_j$ $(j = 1,2,3,4)$ there are at least two vlaues, say $a_1$ and $a_2$, which are not completely multiple values of $f(z)$. Next by Picard's theorem, $f(z)$ takes at least one of the values $a_j$ $(j = 1,2)$, say $a_1$, an infinite number of times. Therefore there is a point $b$ such that

$$f(b) = a_1,\ f'(b) \neq 0.$$

The function $f(z)$ is then univalent in a circle $\gamma:|z-b| < \rho$ and maps $\gamma$ onto a domain $d$ containing the point $a_1$. Accordingly there is a function $g(z)$ which is holomorphic in $d$ and satisfies in $d$ the inequality

$$f\{g(z)\} = z.$$

Consider a circle $C_1:|z-a_1| < r_1$ which belongs to $C \cap d$ and does not contain the point $z_0$. Now suppose that there does not exist a point $z_1 \in C_1 \cap H(f)$. Then evidently the sequence of functions

$$F_n(z) = \frac{f_n(z)-z}{g(z)-z} \quad (n = 1,2,\ldots)$$

are holomorphic and do not take the values 0 and 1 in $C_1$. Consequently the family $\{F_n(z)\ (n = 1,2,\ldots)\}$ is normal in $C_1$. This implies that the family $\{f_n(z)\ (n = 1,2,\ldots)\}$ is also normal in $C_1$. Since $a_1 \in E(f)$, we get a contradiction. Thus there exists a point $z_1$ such that

$z_1 \in C,\ z_1 \neq z_0,\ z_1 \in H(f).$

Hence $z_0$ is a point of accumulation of $H(f)$. The proof of the theorem is now complete.

REFERENCES

[1] Fatou, P., Sur l'itération des fonctions transcendantes entières, Acta Math., 47(1926).

[2] Montel, P., Lecons sur les familles normales de fonctions analytiques et leurs applications, Paris (1927).

[3] Nevanlinna, R., Le théorème de Picard-Borel et la théorie des fonctions méromorphes, Paris (1929).

[4] Hayman, W.K., Meromorphic functions, Oxford (1964).

[5] Rosenbloom, P.C., The fix-points of entire functions, Medd. Lunds Univ. mat. Sem., Suppl-Bd. M. Riesz, 186(1952).

[6] Baker, I.N., Repulsive Fixpoints of Entire Functions, Math. Z. 104 (1968).

DEPARTMENT OF MATHEMATICS
PEKING UNIVERSITY
Beijing

Contemporary Mathematics
Volume 48, 1985

# ALGEBROID FUNCTIONS AND THEIR APPLICATIONS TO ORDINARY DIFFERENTIAL EQUATIONS

Yu-zan He

## I. INTRODUCTION

Starting in 1956, Hiong [13] was the first one in China who engaged in studying algebroid functions. He said in [14], that Poincaré first introduced the algebroid functions. Darboux believed that it is an important class of functions. Later Painlevé encountered them in ordinary differential equations. Hiong proposed the study of the algebroid functions, the applications of Nevanlinna's theory to ordinary differential equations and quasi-entire functions. The study was then joined by Chi-Tai Chuang. Under the guidance and the influence of Hiong and Chuang, some of their students have continued to study these subjects and have obtained a series of results. In the present article we shall give some account of them. In Sections II-III, we give some results on the algebroid functions, such as the generalizations of the second fundamental theorem, deficient values, uniqueness theorem and Borel direction. In Sections IV-V, we give some applications of algebroid functions to ordinary differential equations, such as the generalization of Malmquist's theorem, the value distribution property and the growth of algebroid solutions of ordinary differential equations.

## II. GENERALIZATIONS OF THE SECOND FUNDAMENTAL THEOREM TO ALGEBROID FUNCTIONS

An analytic function $w(z)$ is called an algebroid function, if it satisfies the following functional equation

$$\psi(z,w) \equiv A_{\nu}(z)w^{\nu} + A_{\nu-1}(z)w^{\nu-1} + \ldots + A_0(z) = 0 \tag{1}$$

where $A_j(z)$ $(j = 0,1,\ldots,\nu)$ are holomorphic functions in $D_R = \{z, |z| < R\}$. If $z$ is not a branch point, then (1) determines $\nu$ values of $w$, so $w(z)$ is a $\nu$-valued function. Soon after R. Nevanlinna constructed the value distribution theory of meromorphic functions, G. Valiron [23], E. Ullrich [22] and H. Selberg [21] generalized Nevanlinna's theory to the algebroid functions.

Let us denote by $L$ a curve joining all branch points of $w(z)$, then the determinations $\{w_j(z)\}$ of $w(z)$ are simple-valued functions in $D_R \setminus L$, the characteristic function is defined by

$$T(r,w) = m(r,w) + N(r,w)$$

where

$$m(r,w) = \frac{1}{\nu}\sum_{j=1}^{\nu}\frac{1}{2\pi}\int_0^{2\pi}\log^+|w_j(re^{i\theta})|d\theta$$

$$m(r,\frac{1}{w-a}) = \frac{1}{\nu}\sum\frac{1}{2\pi}\int_0^{2\pi}\log^+\frac{1}{|w_j(re^{i\theta})-a|}d\theta, \quad a \in C$$

$$N(r,w) = \frac{1}{\nu}N(r,\frac{1}{A\nu})$$

$$N(r,\frac{1}{w-a}) = \frac{1}{\nu}N(r,\frac{1}{\psi(z,a)}), \qquad a \in C$$

we have the first fundamental theorem.

Let $w(z)$ be an algebroid function defined by (1), then

$$m(r,\frac{1}{w-a}) + N(r,\frac{1}{w-a}) = T(r,w) + \frac{1}{\nu}\log\left|\frac{A_\nu(0)}{\psi(0,a)}\right| + \varepsilon(r,a)$$

where

$$|\varepsilon(r,a)| \le \log^+|a| + \log 2.$$

For the logarithmic derivative of an algebroid function, we have [23], [6] the following result. Let $w(z)$ be an algebroid function defined by (1), then

$$M(r,\frac{w^{(k)}}{w}) < \alpha_k\log^+T(t,w) + \beta_k\log\frac{t}{t-r} + \gamma_r\log^+\frac{1}{r} + \delta_k;\ t > r,$$

where $\alpha_k$, $\beta_k$ and $\gamma_k$ are some constants depending on $k$, $\delta_k$ is a constant depending on $k$, $\mu$ and the behavior of $\{A_j(z)\}j = 0,1,\dots,\nu$ at $z = C$.

The second fundamental theorem on the algebroid functions was proved by Valiron, Ullrich and Selberg in different ways.

Let $w(z)$ be a $\nu$-valued algebroid function and $a_j \in \overline{C}(j = 1,2,\dots,p)$ be $p$ different numbers, then

$$(p-2\nu)T(r,w) \le \sum_{j=1}^{p} N(r,\frac{1}{w-a_j}) - N_1(r) + S(r) \tag{2}$$

where $N_1(r)$ is the counting function for all multiple values of $w(z)$ and $S(r)$ is the remainder term.

It is well known that Milloux obtained a second fundamental theorem on meormorphic functions involving the derivatives. In [14], Hiong extended the theorem to the algebroid functions. He proved

THEOREM 1. Let $w(z)$ be a $\nu$-valued algebroid functions, $a_j \in C(j= 1,2,\dots,p)$

be $p$ distinct finite complex number and $b_k \in \overline{C}$, $b_k \neq 0$ $(k = 1,2,\ldots,q)$ be $q$ distinct numbers, then

$$p(q-2\nu)T(r,w) \leq (q-2\nu) \sum_{j=1}^{p} N_\nu(r, \frac{1}{w-a_j}) + \sum_{k=1}^{q} \overline{N}(r, \frac{1}{w'-b_k}) + S(r) \tag{3}$$

where $N_\nu(r, \frac{1}{w-a})$ is the counting function for the zeros of $w(z) - a$ in which the multiplicity is counted if the multiple order $\tau \leq \nu$, and only $\nu$ times if $\tau > \nu$, $\overline{N}(r, \frac{1}{w'-b})$ is the counting function for the distinct zeros of $w'(z) - b$.

In [7], He gave another version of the second fundamental theorem involving the derivative which extended one of Hiong's theorems on the meromorphic fnctions. He proved

THEOREM 2. Let $w(z)$ be a $\nu$-valued algebroid function, $a_j$ $(j=1,2,\ldots,p)$ and $b_k$ $(k=1,2,\ldots,q)$ be two groups of complex numbers which are finite, nonzero and distinct in each group, then

$$(p+q-6(\nu-1))T(r,w) \leq (q+1)N_\nu(r, \frac{1}{w}) + 2\overline{N}(r, \frac{1}{w}) + \sum_{j=1}^{p} N_\nu(r, \frac{1}{w-a_j})$$
$$+ N_\nu(r, \frac{1}{w'}) + \sum_{k=1}^{q} N_\nu(r, \frac{1}{w'-b_k}) + S(r). \tag{4}$$

Because the coefficients of the terms of the summation of the counting functions are equal to one, (4) can be used to discuss some uniqueness theorems.

## III. DEFICIENT VALUES, MULTIPLE VALUES, UNIQUENESS THEOREM AND BOREL DIRECTION

For any complex number $\alpha \in \overline{C}$, we define the deficiency and the total ramification of $\alpha$ as follows

$$\delta(\alpha) = 1 - \overline{\lim_{r\to R}} \frac{N(r, \frac{1}{w-\alpha})}{T(r,w)}$$

$$\theta(\alpha) = 1 - \lim_{i\to R} \frac{\overline{N}(r, \frac{1}{w-\alpha})}{T(r,w)} .$$

From the first fundamental theorem it follows

$$0 \leq \delta(\alpha) \leq \theta(\alpha) \leq 1$$

In the case $R = +\infty$, from the second fundamental theorem it follows

$$\sum_{\alpha\in\overline{C}} \delta(\alpha) \leq \sum_{\alpha\in\overline{C}} \theta(\alpha) \leq 2\nu.$$

In [5], He gave a second fundamental theorem in its precise form.

By using it he obtained some theorems on the deficient values and the uniqueness theorems.

Suppose $\gamma(\geq -1)$ is an integer, $\overline{N}^{\gamma)}(r, \frac{1}{w-a})$ denotes the counting function of the zeros of $w(z) - a$ in which a multiple zero is counted only once if its multiple order $\tau \leq \nu$, and is not counted if $\tau > \nu$, then

$$(\gamma p - 2\nu)(\gamma+1))T(r,w) \leq \gamma \sum_{j=1}^{p} \overline{N}^{\gamma)}(r, \frac{1}{w-a_j}) + S_\gamma(r) \tag{5}$$

Let

$$\theta_{\gamma)}(\alpha) = 1 - \overline{\lim_{r\to R}} \frac{\overline{N}^{\gamma)}(r, \frac{1}{w-\alpha})}{T(r,w)}.$$

He proved the following theorem [8]:

THEOREM 3. Let $w(z)$ be a $\nu$-valued algebroid function in $D_R$. Then

(i) in the case $R = +\infty$

$$\sum_{\alpha\in\overline{C}} \theta_{\gamma)}(\alpha) \leq \frac{2\nu(\gamma+1)}{\gamma}$$

(ii) in the case $R = 1$, if $T(r,w)$ satisfies $\overline{\lim_{r\to 1}} \frac{T(r,w)}{\log\frac{1}{1-r}} = \infty$, then (6) holds. If

$$\sum_{\alpha\in\overline{C}} \theta_{\gamma)}(\alpha) \geq \ell > \frac{2\nu(\gamma+1)}{\gamma}$$

then

$$\overline{\lim_{r\to 1}} \frac{T(r,w)}{\log\frac{1}{1-r}} \leq \frac{\gamma+1}{\gamma\ell - 2\nu(\gamma+1)}.$$

If $\theta_{\gamma)}(\alpha) = 1$, we call $\alpha$ a full deficient value with order $\gamma$. From (6) it follows that a $\nu$-valued algebroid function at most $4\nu$, $3\nu$, $2\nu+2$ or $2\nu$ full deficient values with order 1, 2, $\nu$ or $2\nu+1$ can exist respectively.

Let us say that a value $\alpha$ has multiplicity at least $m$, if all the roots of the equation $w(z) = \alpha$ have multiplicity at least $m$. From (2) it follows

$$\sum_{j=1}^{p} (1 - \frac{1}{m_j}) \leq 2\nu, \tag{7}$$

especially if $m_j \geq 2$. Then there are $4\nu$ such values $a_j$ at most.

Under what circumstances can two different algebroid functions $w(z)$, $\hat{w}(z)$ take the same value at the same points? In [5], He proved the following uniqueness theorem.

THEOREM 4. Let $w(z)$ be a $\nu$-valued algebroid function and $\overline{E}^{\gamma)}(a)$ denotes the set of a-points of $w(z)$ whose multiplicity $\tau \le \gamma$ each being counted only once. Then $w(z)$ is determined uniquely by $p = 4 + 1 + [\frac{2\nu}{\gamma}]$ sets $\overline{E}^{\gamma)}(a_j)$ $(j = 1,2,\dots,p)$, where $[\alpha]$ denotes the integer part of $\alpha$. Especially if $\gamma = 1, 2, \nu$ and $2\nu + 1$, then $p = 6\nu + 1$, $5\nu + 1$, $4\nu + 3$ and $4\nu + 1$ respectively.

We sketch the proof. Let $\hat{w}(z)$ be an algebroid function defined by

$$\Phi(z,\hat{w}) \equiv B_\mu(z)\hat{w}^\mu + B_{\mu-1}(z)\hat{w}^{\mu-1} + \dots + B_0(z) = 0.$$

Without loss of generality, we suppose $\nu \ge \mu$, and set

$$\overline{N}_0^{\gamma)}(r,a) = \frac{\mu+\nu}{2\mu\nu}\left(\int_0^r \frac{\overline{n}_0^{\gamma)}(t,a) - \overline{n}_0^{\gamma)}(0,a)}{t}\,dt + \overline{n}_0^{\gamma)}(0,a)\log r\right)$$

and

$$\overline{N}_{12}^{\gamma)}(r,a) = \overline{N}^{\gamma)}(r, \frac{1}{w-a}) + \overline{N}^{\gamma)}(r, \frac{1}{\hat{w}-a}) - 2\,\overline{N}_0^{\gamma)}(r,a)$$

where $\overline{n}_0^{\gamma)}(t,a)$ denotes the number of the common roots of $w(z) = a$ and $\hat{w}(z) = a$ containing in $|z| < t$, whose multiple order $\tau \le \nu$ and each being counted only once.

Applying (5) to $w(z)$ and $\hat{w}(z)$, and assuming $\nu \ge \mu$, we get

$$(\gamma p - 2\nu(\gamma+1))(T(r,\hat{w}) + T(r,w)) \le \gamma \sum_{j=1}^{p} \overline{N}_{12}^{\gamma)}(r,a_j) + 2\gamma \sum_{j=1}^{p} \overline{N}_0^{\gamma)}(r,a_j) + S_\gamma(r).$$

Suppose $w(z) \not\equiv \hat{w}(z)$, then the discriminant

$$R(\psi,\Phi) \equiv (A_\nu(z))^\nu (B_\mu(z))^\nu \prod_{\substack{1\le j\le\nu \\ 1\le k\le\mu}} (w_j(z) - \hat{w}_k(z))^2 \not\equiv 0$$

and

$$\frac{2\mu\nu}{\mu+\nu} \sum_{j=1}^{p} \overline{N}_0^{\gamma)}(r,a_j) \le N(r, \frac{1}{R(\Psi,\Phi)}) \le \mu\nu(T(r,w) + T(r,\hat{w})) + O(1)$$

hence

$$(\gamma p - 2\nu(2\gamma+1))(T(r,w) + T(r,\hat{w})) \le r \sum_{j=1}^{p} \overline{N}_{12}^{\gamma)}(r,a_j) + S_\gamma(r).$$

Similar to the proof of uniqueness theorem on the meromorphic functions, Theorem 4 follows.

By using Theorem 2, He proved another uniqueness theorem which shows that an algebroid function can be determined uniquely by $4\nu + 1$ sets taken from $4\nu$ sets of a-points of $w(z)$ and $4\nu$ sets of b-points of $w'(z)$.

If $A_j(z)(j = 0,1,\ldots,\nu)$ are linearly independent. He gave a second fundamental theorem in the following form

$$(\sum_{j=1}^{p} \frac{\gamma_j-\nu+1}{\gamma_j+1} - \nu - 1)T(r,w) \leq \sum_{j=1}^{p} \frac{\gamma_j-\nu+1}{\gamma_j+1} N^{\gamma_j)}(r, \frac{1}{w-a_j}) + S(r) \tag{9}$$

where $\gamma_j(\nu-1)$ are some integers and $N_\nu^{\gamma)}(r, \frac{1}{w-a})$ denotes the counting function of the zeros of $w(z) - a$, in which the multiple zeros being counted $\tau$ times if the multiple order $\tau \leq \nu$, only $\nu$ times if $\nu + 1 \leq \tau \leq \nu$, and not counted if $\tau > \gamma$.

On the growth of the algebroid functions with several deficient values, Gao [4] proved[1)]

THEOREM 5. Let $w(z)$ be a $\nu$-valued transcendental algebroid function with $\nu + 1$ deficient values. Then the lower order of $w(z)$ is positive.

In [30], Zhao considered the inverse function of the entire algebroid function with 2 branches. In [16], Lü discussed the relationship between the growth order and the direct transcendental singularities of the inverse function. (See the article "Distribution of the values of meromorphic functions of this book.)

Lü and Gao [18] proved the existence of the Borel direction for an algebroid function of finite positive order

THEOREM 6. Let $w(z)$ be a $\nu$-valued algebroid function of order $\rho$; $0 < \rho < +\infty$, $0 < \rho < \infty$, then there exists a direction $L = \{z \in \mathbb{C}, \arg z = \theta_0\}$ such that for any given $\delta$ $(0 < \delta < \pi/2)$

$$\overline{\lim_{r\to\infty}} \frac{\log^+ n(r,\Delta(\theta_0,\delta),a)}{\log r} = \rho$$

for any $a \in \overline{C}$ with $2\nu$ possible exceptions, where $n(r,\Delta(\theta_0,\delta),a)$ is the number of a-points of $w(z)$ in $\Delta(\theta_0,\delta) = \{z, |\arg z - \theta_0| < \delta\} \cap \{z, |z| < r\}$. (See "Distribution of values of meromorphic functions".)

## IV. GENERALIZATION OF MALMQUIST'S THEOREM

The existence problem on meromorphic or algebroid solutions of some first order ordinary differential equations was first investigated by J. Malmquist, [19]. He proved the following important theorem which bears his name: If the ordinary differential equation with rational coefficients

$$\frac{dw}{dz} = \sum_{k=0}^{p} a_k(z)w^k / \sum_{j=0}^{q} b_j(z)w^j \tag{10}$$

1) Theorem 5 was proved independently by M. Ozawa in [20].

has a transcendental meromorphic solution. Then $q = 0$ and $p \le 2$.

In 1933, K. Yosida [28] presented an elegant proof and an extension of Malmquist's theorem by using the Nevanlinna's theory. Starting in 1950, a systematic study of the implications of Nevanlinna's theory for ordinary differential equations had been undertaken by H. Wittich [24]. A series of further important results were obtained by mathematicians in Japan, Germany, Soviet Union, America, Finland and China.

In many circumstances the differential equations have algebroid solutions, for example, the linear differential equation $\frac{dw}{dz} + \frac{1}{2}(\cot z)w = 0$ has the 2-valued algebroid solution $w(z) = \frac{1}{\sin z}$. He and Xiao concentrated their attention on algebroid solutions. They obtained some generalizations of Malmquist's theorem in precise form and gave some examples which show that the bounds in their theorem are the best possible.

Let

$$\Omega(z,w) = R(z,w) \tag{11}$$

where $\Omega(z,w) \equiv \sum a_{(i)}(z)(w)^{i_0}(w')^{i_1}\dots(w^{(n)})^{i_n}$ is a differential polynomial with meromorphic coefficients $\{a_{(i)}(z)\}$, and $R(z,w) \equiv P(z,w)/Q(z,w) \equiv \sum_{k=0}^{p} a_k(z)w^k / \sum_{j=0}^{q} b_j(z)w^j$ is a rational function in $w$ with meromorphic coefficients.

We first give a definition of an admissible solution as follows.

Let $w(z)$ be an algebroid solution of (11) and $U(r) = r^{\rho(r)}$ be a type function of $T(r,w)$, where $\rho(r)$ is a precise order of Valiron (or Chuang) if $U(r)$ is of finite order (or of infinite order). Then $w(z)$ is an admissible solution, if it satisfies

$$\overline{\lim_{r\to\infty}} \frac{S_1(r)}{T(r,w)} = 0$$

outside a possible exceptional set $E$ with a finite linear measure, where

$$S_1(r) = \sum T(r,a_{(i)}) + \sum_{k=0}^{p} T(r,a_k) + \sum_{j=0}^{q} T(r,b_j).$$

He and Xiao proved([11], [27])

THEOREM 7. If the equation (11) admits at least one admissible $\nu$-valued algebroid solution, then

$$q \le 2\sigma(\nu-1) \quad \text{and} \quad p \le \mathrm{Min}\{2\sigma(\nu-1)+\lambda+\overline{\mu},\ q+\lambda+\overline{\mu}\nu\}, \tag{12}$$

where $\lambda = \mathrm{Max}\{\sum_{\lambda=0}^{n} i_\alpha\}$, the maximum is taken over all the index $\{i_0, i_1, \ldots, i_n\}$ of the terms in $\Omega(z,w)$, $\bar{\mu} = \mathrm{Max}\{\sum_{\alpha=1}^{n} \alpha_{i\alpha}\}$, $\sigma = \mathrm{Max}\{\sum_{\alpha=1}^{n} (2\alpha-1)i_\alpha\}$.

In order to prove Theorem 7, it is necessary to give a precise estimation for the characteristic function of $\Omega(z, w(z))$ and of the composite function $F(z,w(z))$. We have

LEMMA 1. If the algebroid function $w(z)$ satisfies

$$P(z,w(z))\Omega(z,w(z)) = \hat{\Omega}(z,w(z))$$

where

$$P(z,w) = \sum_{k=0}^{p} a_k(z)w^k, \quad \hat{\Omega}(z,w) = \sum b_{(j)}(z)(w)^{j_0}\ldots(w^{(m)})^{j_m} \quad \text{and}$$

$$p \geq \hat{\lambda} = \mathrm{Max}\{\sum_{\beta=0}^{m} j_\beta\},$$

then

$$m(r,\hat{\Omega}(z,w(z))) = S_1(r) + S(r)$$

where

$$S_1(r) = K\{\sum T(r,a_{(i)}) + \sum T(r,b_{(j)}) + \sum T(r,a_k)\}.$$

It is well known that the poles of the derivative $w^{(k)}(z)$ of an algebroid function $w(z)$ arise not only from the poles of $w(z)$, but also from some branch points. He proved [5]

$$N(r,w^{(k)}) \leq N(r,w) + k\bar{N}(r,w) + (2k-1)N_x(r,w) \leq N(r,w) + k\bar{N}(r,w) +$$
$$2(\nu-1)(2k-1)T(r,w) + O(1)$$

where $N_x(r,w)$ is the counting function of the branch points of $w(z)$. From the above inequality we get the following result.

LEMMA 2. Let $w(z)$ be a $\nu$-valued algebroid function, then

$$T(r,\Omega(z,w(z))) \leq \bar{\mu}\bar{N}(r,w) + (\lambda+2\sigma(\nu-1))T(r,w) + S_1(r) + S(r). \tag{13}$$

For the characteristic function of the composite function $R(z,w(z))$, we have

LEMMA 3. Let $w(z)$ be an algebroid function, then

$$T(r,R(z,w(z))) = \mathrm{Max}\{p,q\}T(r,w) + S_1(r). \tag{14}$$

We shall only present a sketch proof of Theorem 7. We first write (11) as follows

$$\Omega(z,w) = P_1(z,w) + P_2(z,w)/Q(z, w),$$

then (11) becomes

$$Q(z,w)\hat{\Omega}(z,w) = P_2(z,w)$$

where $\hat{\Omega}(z,w) = \Omega(z,w) - p_1(z,w)$ and the degree of $P_1(z,w)$ in $w$ is $p-q$, i.e., $\deg P_1 = p-q$ and $\deg P_2 < q$. By Lemma 1

$$m(r,\hat{\Omega}(z,w(z))) = S_1(r) + S(r).$$

For the counting function of the poles of $\hat{\Omega}(z,w(z))$, we get

$$N(r,\hat{\Omega}(z,w(z))) \le 2\sigma(\nu-1)T(r,w) + S_1(r) + S(r)$$

thus

$$T(r,\hat{\Omega}(z,w(z))) \le 2\sigma(\nu-1)T(r,w) + S_1(r) + S(r).$$

On the other hand, by Lemma 3

$$T(r,\hat{\Omega}) = T(r,P_2/Q) = qT(r,w) + S_1(r)$$

thus

$$qT(r,w) \le 2\sigma(\nu-1)T(r,w) + S_1(r) + S(r).$$

Noting the definition of admissible solution, the first inequality of (12) follows.

To prove the second assertion of Theorem 7, we first write (11) as follows

$$P(z,w) = \Omega(z,w)Q(z,w).$$

By Lemma 2, we have

$$T(r,\Omega Q) \le (q+\lambda)T(r,w) + \bar{\mu}\nu\bar{N}(r,w) + S_1(r) + S(r).$$

By virtue of Lemma 3 and the above inequality,

$$\begin{aligned} pT(r,w) &= T(r,P(z,w(z))) + S_1(r) = T(r,Q(z,w(z))\Omega(z,w(z))) + S_1(r) \\ &\le (q+\lambda)T(r,w) + \bar{\mu}\nu\bar{N}(r,w) + S_1(r) + S(r) \end{aligned}$$

hence

$$p \le q + \lambda + \bar{\mu}\nu.$$

On the other hand, He and Xiao proved [27]

$$\operatorname{Max}\{p,q\} \le \lambda + \bar{\mu} + 2\sigma(\nu-1).$$

It yields immediately the second inequality of (12).

The following example shows that the bounds in Theorem 7 can be attained.

EXAMPLE. It is easy to show that the algebroid function defined by

$$p(z)w^2 + \tan z + q(z) = 0$$

is an admissible solution of the equation

$$\frac{dw}{dz} = \frac{p^2(z)w^4+p'(z)w^3+(1+p(z)q(z))w^2+q'(z)w+q^2(z)}{q(z)-p(z)w^2}$$

where $p(z)$ and $q(z)$ are two polynomials in $z$. Since $\lambda = \overline{\mu} = \sigma = 1$, $q = 2$ and $p = 4$, by Theorem 7, the upperbounds of $q$ and $p$ are $2\sigma(\nu-1) = 2$ and $2\sigma(\nu-1) + \lambda + \overline{\mu} = 4$ respectively. It shows that the bounds in Theorem 7 are the best possible.

## V. VALUE DISTRIBUTION PROPERTY AND GROWTH OF ALGEBROID SOLUTIONS

We first have

THEOREM 8. Let $w(z)$ be an admissible $\nu$-valued algebroid solution of (11), if $p > q + \lambda$. Then

$$\delta(\infty) = 1 - \varlimsup_{r\to\infty} \frac{N(r,w)}{T(r,w)} = 0 \quad \text{and} \quad \theta(\infty) = 1 - \varlimsup_{r\to\infty} \frac{\overline{N}(r,w)}{T(r,w)} \le 1 - \frac{p-(q+\lambda)}{\overline{\mu}\nu}.$$

On the growth of algebroid solution, we obtained some results which extended some results on the meromorphic solutions obtained by S. Bank and I. Lanie. The following theorem (Theorem 9) summarizes such results. Let

$$\Omega(z,w) = \sum a_{(i)}(z)(w)^{i_0}(w')^{i_1}\cdots(w^{(n)})^{i_n} = 0 \tag{15}$$

be an algebraic differential equation with meromorphic coefficients and $\Omega_\ell(z,w)$ denote a homogeneous part of $\Omega(z,w)$ of degree $\ell$. $A_\ell(z)$ denote the sum of all coefficients $a_{(i)}(z)$ in $\Omega_\ell(z,w)$ having multi-indices of maximal weight $\overline{\mu}_{(i)} = \sum_{\alpha=1}^{n} \alpha i_\alpha$, i.e., for $k = \operatorname{Max}\{\overline{\mu}_{(i)}\}$, we have

$$A_\ell(z) = \sum_{\substack{\lambda_{(i)}=\ell,\\ \overline{\mu}_{(i)}=k}} a_{(i)}(z).$$

He and Laine proved

THEOREM 9. Let $w(z)$ be an algebroid solution of (15).

(i) If $w(z)$ does not satisfy all homogeneous equations $\Omega_\ell(z,w) = 0$, then for every $\sigma > 1$, there exist constants $K$ and $r_0 \ge 1$ such that for all $r \geq r_0$

$$T(r,w) \leq KF(\sigma r)$$

where

$$F(r) = \overline{N}(r,w) + \overline{N}(r, \frac{1}{w}) + N_x(r,w) + S_1(r)$$

$$S_1(r) = \sum T(r,a_{(i)}).$$

(ii) If $w(z)$ also satisfies all $\Omega_\ell(z,w) = 0$, and if when $\Omega_\ell(z,w) \not\equiv 0$ for some $\ell$, we have $A_\ell(z) \not\equiv 0$, then for any $\sigma > 1$, there exist constants $K$, $K_1$ and $r_0 \geq 1$ such that for all $r \geq r_0$

$$T(r,w) \leq Kr^2 \exp\{K_1 F(\sigma r) \log(rF(\sigma r))\}.$$

COROLLARY. Let $w(z)$ be an algebroid solution of a linear differential equation

$$\sum_{i=0}^{n} f_j(z) w^{(j)} = 0$$

then

$$T(r,w) \leq Kr^2 \exp\{K_1 F_1(\sigma r) \log(rF_1(\sigma r))\}$$

where

$$F_1(r) = \overline{N}(r, \frac{1}{w}) + N_x(r,w) + S_1(r).$$

To prove Theorem 9, He and Laine [12] gave the following lemma which extended a result of Bank's [1].

LEMMA 4. Let $w(z)$ be an algebroid function and $y(z) = w'(z)/w(z)$, then for any $\alpha > 1$, ther exist constants $A$, $B$ and $r_1 \geq 1$ such that for all $r \geq r_1$

$$T(r,w) \leq A\{rN(\alpha r,w) + r^2 \exp BT(\alpha r,y) \log(rT(\alpha r,y)))\}.$$

Theorem 9 shows that the growth of an algebroid solution can be estimated uniformly in terms of the growth of the coefficients and the counting functions of the zeros, poles and branch points of $w(z)$.

## REFERENCES

[1] Bank, S., On determining the growth of meromorphic solutions of algebraic differential equations having arbitrary entire coefficients, Nagoya Math. J. 49 (1973), 53-65.

[2] Chuang Chi-tai, Singular directions of meromorphic functions, Science Press Beijing, China 1982.

[3] Gackstatter, F. and Laine, I., Zur theorie der gewöhnlichen differentialgleichungen im komplexen, Ann. Polon. Math. 38 (1980), 260-287.

[4] Gu Yongxing, The growth of algebroid functions with several deficient values, Contemporary Math., 25 (1983), 45-49.

[5] He Yuzan, Sur un problème d'unicité relatif aux fonctions algébroides, Scientia Sinica, 14 (1965), 174-180.

[6] He Yuzan, Sur les fonctions algébroides et leurs dérivées (I), Acta Math. Sinica, 15 (1965), 281-295.

[7] He Yuzan, Sur les fonctions algébroides et leurs dérivées (II), Acta Math. Sinica, 15 (1965), 500-510.

[8] He Yuzan, On the multiple values of algebroid functions, Acta Math. Sinica, 22 (1979), 733-742.

[9] He Yuzan, On algebroid solutions of ordinary differential equations, Acta Math. Sinica 24 (1981), 464-471.

[10] He Yuzan and Xiao Xiuzhi, On Malmquist's theorem for algebraic differential equations, Kexue Tongbao, 28 (1983), 165-169.

[11] He Yuzan and Xiao Xiuzhi, Admissible solutions of ordinary differential equations, Contemporary Math., 25 (1983), 51-60.

[12] He Yuzan and Laine, I., On the growth of algebroid solutions of ordinary differential equations, Math. Scand. (to appear).

[13] Hiong King-lai, Sur la croissance des fonctions algébroides en rapport avec leurs dérivées, C.R. Acad. Sc., 241 (1956), 3032-3035.

[14] Hiong King-lai, Sur les fonctions meromorphes et les fonctions algébroides. Extensions d'un théorème de M. R. Nevanlinna, Gauthier-villar, Paris, 1957.

[15] Hiong King-lai, Some achievements of the theory on meromorphic functions, Advances in Math. 6 (1963), 307-320.

[16] Lü Yinian, On direct transcendental singularities of the inverse function of an algebroidal function, Scientia Sinica, 23 (1980), 407-415.

[17] Lü Yinian, On Borel direction of an algebroidal function, Scientia Sinica Series A, 25 (1982), 25-30.

[18] Lü Yinian and Gu Yongxing, On the existence of Borel direction of an algebroidal function, Kexue Tongbao, 28 (1983), 264-266.

[19] Malmquist, J., Sur les fonctions a un numbre fini des branches définies par les equations differentielles du premier ordre, Acta Math. 36 (1913), 297-343.

[20] Ozawa, M., On the growth of algebroid functions with several deficiencies (II), Kodai Math. Sem. Rep. 22 (1970), 129-137.

[21] Selberg, H., Über die wertverteilung der algebroiden funktionen, Math. Z., 31 (1930), 709-728.

[22] Ullrich, E., Über den Einfluss der verzweigtheit einer algebroide auf ihre wertverteilung, J. Reine Ang. Math., 167 (1931), 198-220.

[23] Valiron, G., Sur la dérivée des fonctions algebroides, Bull. Soc. Math. Fr. 59 (1929), 17-39.

[24] Wittich, H., Neuere untersuchungen über eindeutige analysche funktionen, Springer-Verlag, Berlin-Gottingen-Heidelberg, 1955.

[25] Xiao Xiuzhi, The generalization of Malmquist's theorem on a class of ordinary differential equations, Acta Math., Sinica, 15 (1965), 397-405.

[26] Xiao Xiuzhi, On the growth of single-valued solutions of linear differential equations, Wuhan University Journal (National Science Edition), 1979, No. 2.

[27] Xiao Xiuzhi and He Yuzan, Meromorphic and algebroid solutions of higher-order algebraic differential equations, Scientia Sinica (series A), 26 (1983), 1034-1043.

[28] Yosida, K., A generalization of Malmquist's theorem, Japan, J. Math., 9 (1933), 253-256.

[29] Yosida, K., On algebroid solutions of ordinary differential equations, Japan J. Math., 10 (1934), 1-20.

[30] Zhao Jing-yi, On entire algebroid functions with 2 branches, Math. Cб. 37 (79) (1955), 573-576.

Institute of Mathematics
Academia Sinica
Beijing

Contemporary Mathematics
Volume 48, 1985

# SUMMARY OF RECENT RESEARCH ACCOMPLISHMENTS IN VALUE DISTRIBUTION THEORY AT EAST CHINA NORMAL UNIVERSITY

R.F. Lee, C.J. Dai, G.D. Song

In this paper we shall report the main results in value distribution theory which have been obtained since 1979, by the function theory group at the Mathematics Department of East China Normal University. The research topics covered include investigations on Borel direction, Nevanlinna direction, number of deficiencies of a meromorphic function and its derivatives, deficient functions, normal family and factorization theory, etc.

## PART A. Definitions and notations

Let $f(z)$ be a meromorphic function in the plane. Denote

$$m(r,f) = m(r,\infty) = \frac{1}{2\pi}\int_0^{2\pi} \log^+|f(re^{i\theta})|\,d\theta,$$

$$m(r,a) = m(r,\frac{1}{f-a}),$$

$$N(r,f) = N(r,\infty) = \int_0^r \frac{n(t,f)-n(0,f)}{t}\,dt + n(0,f)\log r,$$

$$N(r,a) = N(r,\frac{1}{f-a}),$$

where $a$ is a finite complex number, $n(r,f)$ denotes the number of poles of $f(z)$ in $|z| \le r$.

Denote

$$T(r,f) = m(r,f) + N(r,f),$$

$$T(r,a) = T(r,\frac{1}{f-a}).$$

$T(r,f)$ is called the characteristic function of $f(z)$.

For a complex number $a$ $(|a| \le \infty)$, the deficiency $\delta(a,f)$ is defined by

$$\delta(a,f) = 1 - \overline{\lim_{r\to\infty}}\frac{N(r,a)}{T(r,f)}.$$

If $\delta(a,f) > 0$, then $a$ is called a deficient value of $f(z)$.

The order $\rho$ and lower order $\mu$ of $f(z)$ are defined by

$$\rho = \overline{\lim_{r\to\infty}} \frac{\log T(r,f)}{\log r}, \quad \mu = \underline{\lim_{r\to\infty}} \frac{\log T(r,f)}{\log r}.$$

Let $f(z)$ be a meromorphic function of order $\rho$ $(0 < \rho < \infty)$. Valiron [1] proved in 1928 that there exists at least a direction B: $\arg z = \theta$, such that for any three distinct complex numbers $a,b,c$ (one of them may be $\infty$), we have

$$\overline{\lim_{r\to\infty}} \frac{\log\{n(r,\theta,\varepsilon,f=a)+n(r,\theta,\varepsilon,f=b)+n(r,\theta,\varepsilon,f=c)\}}{\log r} = \rho, \tag{1}$$

where $n(r,\theta,\varepsilon,f=A)$ denotes the number of roots of the equation $f(z) - A = 0$ in the sector $\{|z| \le r,\ |\arg z - \theta| \le \varepsilon\}$. The direction B is called a Borel direction of $f(z)$.

PART B. Main results

We shall state our research accomplishments and related known results in the following six aspects.

## I. THE GROWTH OF A FUNCTION ALONG RADIAL LINE AND BOREL DIRECTION

Let $f(z)$ be an entire function of order $\rho$ $(0 < \rho < \infty)$. If, for a certain fixed value $\theta_0$

$$\overline{\lim_{r\to\infty}} \frac{\log^+\log^+|f(re^{i\theta_0})|}{\log r} = \rho_0 \quad (\le \rho). \tag{2}$$

We call $L_{\theta_0}: \arg z = \theta_0$ a radial line of order $\rho_0$ of $f(z)$.

The radial line of $\rho_0$ fill up some angular domains, these angular domains have been called angular domain of order $\rho_0$ (probably it has been degenerated as a radial line).

Dai Chongji [40] proved

THEOREM 1. Let $f(z)$ be an entire function of order $\rho$, for $\rho_0$, $0 \le \rho_0 < \rho$, then number of angular domains of order $\rho_0$ is less than $4\rho$.

Dai Chongji and Ji Shanyu [41] proved $f(z)$ will always be denoted as, unless specifically described, an entire function of order $\rho$ $(0 < \rho < \infty)$.

THEOREM 2. There exists at least one angular domain of order $\rho$ of $f(z)$. Every such angular domain has its opening no less than $\frac{\pi}{\rho}$. All the Borel direction of $f(z)$ lie either inside the angular domain or on its boundary.

THEOREM 3. For every $\theta$, $0 \le \theta < 2\pi$. We have

$$\overline{\lim_{r\to\infty}} \frac{\log^+\log^+|f(re^{i\theta})|}{\log r} = \overline{\lim_{r\to\infty}} \frac{\log^+\log^+|f'(re^{i\theta})|}{\log r} .$$

THEOREM 4. Let $p$ be the number of angular domain of order $\rho$, $q$ be the number of the Borel directions.

(1) If $p < 2\rho$, then $q \geq p+1$.

(2) If $p+1 < 2\rho$ and $q = p+1$, then the angle between every two Consecutive Borel directions is equal to $\frac{\pi}{\rho}$ with one exception.

Earlier Zhang Guanghon [42] proved.

THEOREM A. Let $f(z)$ be a meromorphic function of fine order, if $\infty$ is its Borel exception value, then $f(z)$ and $f^{(n)}(z)$ $(n=1,2,...)$ have one Common Borel direction at least.

Dai Chongji and Ji Shanyu [43] proved

THEOREM 5. Let $f(z)$ be as in Theorem A, if $\infty$ is its Borel exceptional value and the order of $f > \frac{1}{2}$, then $f(z)$ and $f^{(n)}(z)$ $(n=1,2,...)$ have at least two Common Borel directions.

Ji Shanyu [44] proved

THEOREM 6. Let $f(z)$ be a meromorphic function of order $\rho$ $(0 < \rho < \infty)$, $B$: $\arg z = \theta$ be a Borel direction of $f(z)$, then $B$ either is Borel direction of $f^{(n)}(z)$ $(n=1,2,...)$, or is Borel direction of $(\frac{1}{f(z)})^{(n)}$ $(n=1,2,...)$.

THEOREM 7. Let $f(z)$ be as Theorem 6, then either $f(z)$ and $f^{(n)}(z)$ $(n=1,2,...)$ have one Common Borel direction at least or $\frac{1}{f(z)}$ and $(\frac{1}{f(z)})^{(n)}$ $(n=1,2,...)$ have one Common Borel direction at least.

Zhang Qinde [45] proved

THEOREM 8. Let $f(z)$ be a meromorphic function of order $\rho$ $(\frac{1}{2} < \rho < \infty)$, then in any angular domain with opening no less than $\frac{\pi}{\rho}$, among $f(z)$ and $f^{(n)}(z)$ $(n=1,2,...)$ either each of them has at least one Borel direction or none of them has one.

## II. DEFICIENT FUNCTION

Let $a(z)$ be a meromorphic function such that

$$T(r,a(z)) = o\{T(r,f)\}.$$

Then $a(z)$ is called a deficient function of $f(z)$, if

$$\delta(a(z),f) = 1 - \overline{\lim_{r\to\infty}} \frac{N(r, \frac{1}{f-a(z)})}{T(r,f)} > 0. \tag{2}$$

a) The relation between the deficiency and Borel direction of $f(z)$.

In 1975, Yang Le and Zhang Guanghou proved

THEOREM B. [2] Let $f(z)$ be a meromorphic function of order $\rho$, $0 < \rho < \infty$. Let $p$ be the number of deficient values of $f(z)$, and $q$ be the number of Borel directions of $f(z)$, then

$$p \leq q.$$

THEOREM C. [3] Let $f(z)$ be an entire function of order $\rho$, $0 < \rho < \infty$. Let $p$ be the number of deficient values of $f(z)$ other than $\infty$, and $q$ be the number of Borel directions of $f(z)$, then

$$p \leq \frac{q}{2}.$$

Suppose that $\rho(r)$ is a proximate order of $f(z)$, $U(r) = r^{\rho(r)}$. If (3) holds with $U(r)$ in place of $T(r,f)$, then $a(z)$ is called a proximate deficient function of $f$. In this case, the deficiency is denoted by $\delta^*(a(z),f)$. Clearly, $\delta^*(a(z),f) \geq \delta(a(z),f)$.

Zhang Qingde and Pang Xuecheng [19] proved the following two theorems.

THEOREM 9. Let $f(z)$ be as in Theorem B. If $f(z)$ has a proximate deficient function $a(z)$, $a(z) \not\equiv \infty$, then $f(z)$ and all its derivatives have at least a common Borel direction.

THEOREM 10. Let $f(z)$ be as in Theorem B. Denote by $P^*$ the number of the proximate deficient function of $f(z)$. Then

$$P^* \leq \min(q_0, q_1, q_2, \ldots\ldots),$$

where $q_i$ denotes the number of Borel direction of $f^{(i)}(z)$, $i = 0,1,2,\ldots, f^{(0)}(z) = f(z)$. Also, if $q^*$ denotes the number of the common Borel directions of $f(z)$ and all its derivatives, and if $f(z)$ has a proximate deficient function which is not identically equal to $\infty$, then

$$P^* \leq q^*.$$

Pang Xuecheng and Ru Ming [20] proved

THEOREM 11. Let $f(z)$ be as in Theorem B, $\delta^*(\infty,f) = 1$. Let $P^*$ be the number of the proximate deficient function $a(z) (\not\equiv \infty)$, and $q^*$ be as in Theorem 10. Then

$$p^* \le \frac{q^*}{2} .$$

THEOREM 12. Let $f(z)$ be as in Theorem C. Let $p^*$ be the number of the proximate deficient function of $f(z)$, $q_i$ $(i = 0,1,2,\ldots)$ be as in Theorem 10. Then

$$p^* \le \frac{1}{2} \min(q_0, q_1, q_2, \ldots).$$

b) on F. Nevanlinna's conjecture.

It is known that a meromorphic function $f(z)$ has at most a countable number of deficient values, and the sum of deficiencies satisfies:

$$\sum_a \delta(a,f) \le 2.$$

F. Nevanlinna [4] posed in 1929 a conjecture that if $\sum_a \delta(a,f) = 2$, then

(i) The order $\rho$ of $f(z)$ is an integral multiple of $1/2$;

(ii) The total number of deficient values of $f(z) \le 2\rho$.

(iii) Every deficiency is an integral multiple of $1/\rho$.

This is one of the most famous conjectures in value distribution theory of meromorphic functions, which was confirmed by Drasin in 1983. However, it is still an open question that whether or not the conjecture holds for deficient functions.

The conclusion that the number of deficient functions of $f(z)$ is at most countable has been proved by Yang Le [5] and the result that $\sum_{a(z)} \delta(a(z),f) \le 2$ was announced by C.F. Osgood in 1983.

Zhuang Qitai [8] proved in 1964

THEOREM D. Let $f(z)$ be an entire function and $a(z)$ $(\not\equiv \infty)$ be any meromorphic function such that

$$T(r,a(z)) = o\{T(r,f)\}.$$

Then

$$\sum_{a(z)} \delta(a(z),f) \le 1.$$

Li Qingzhong and Ye Yasheng [22] proved.

THEOREM 13. Let $f(z)$ be an entire function of lower order $\mu < \infty$, and $a(z)$ be as in Theorem D. If

$$\sum_{a(z)} \delta(a(z),f) = 1,$$

then the order $\rho$ of $f(z)$ is equal to $\mu$, and is a positive integer.

Jing Lu and Dai Chongji proved

THEOREM 14. [23] Same hypotheses as in Theorem 13. If $\nu(f)$ denotes the number of all the deficient functions $a(z)$ $(\not\equiv \infty)$ of $f(z)$, then

$$\nu(f) \le \mu.$$

THEOREM 15. [24] Same hypotheses as in Theorem 13. If $a(z)$ $(\not\equiv \infty)$ is a deficient function of $f(z)$, then $\delta(a(z),f)$ is a multiple of $1/\mu$.

It is evident that the above two theorems give the extension of F. Nevanlinna's conjecture in the case when $f(z)$ is entire.

C) On S.M. Shah's conjecture

In 1976, S.M. Shah [19] posed the following conjecture: suppose that $f(z)$ is an entire function of lower order $\mu < \infty$ satisfying

$$\sum_{a\neq\infty} \delta(a,f) = 1,$$

then

$$\lim_{r\to\infty} \frac{T(r,f)}{\log M(r,f)} = \frac{1}{\pi}. \tag{4}$$

Zhang Qingde and Huang Jue [18] proved the conjecture in 1983. Recently Ling Qiong and Dai Chongji [25] proved

THEOREM 16. Let $f(z)$ be an entire function of lower order $\mu < \infty$ satisfying

$$\sum_{a(z)} \delta(a(z),f) = 1,$$

where $a(z)$ is any meromorphic function such that $a(z) \not\equiv \infty$ and $T(r,a(z)) = o\{T(r,f)\}$. Then (4) remains valid.

d) On the convergence of $\sum \delta^{1/3}(a(z),f)$.

A. Weitsman proved in 1972 the following

THEOREM E. [10] Let $f(z)$ be a meromorphic function of finite lower order, then

$$\sum_{a} \delta^{1/3}(a,f) < \infty.$$

Zhang Qingde [46] proved

THEOREM 17. Let $f(z)$ be an meromorphic function of finite lower order, then

$$\sum_\nu \delta^{1/3}(a_\nu(z),f) < \infty, \quad a_\nu(z) \in A_n.$$

Specially, $f(z)$ also be an entire function, then

$$\sum_\nu \delta^{1/3}(a_\nu(z),f) < \infty, \quad a_\nu(z) \in A_\infty.$$

Where, $A_n = \{a(z) | a(z)$ is a polynomial with, $\deg a(z) \leq n\} \cup \{\infty\}$,

$A = \{a(z) | a(z)$ is a polynomial$\} \cup \{\infty\}$.

Ling Qun [33] proved

THEOREM 18. Let $f(z)$ be an entire function of finite lower order, and $a(z)$ be any meromorphic function satisfying

$$T(r,a(z)) = o\{T(r,f)\}.$$

Then

$$\sum_{a(z)} \delta^{1/3}(a(z),f) < \infty.$$

## III. HAYMAN'S INEQUALITY, NORMAL FAMILY

R. Nevanlinna's second fundamental theorem shows that three counting functions $N(r,a_1)$, $N(r,a_2)$, $N(r,a_3)$ are needed to bound $T(r,f)$. In 1959, Hayman proved that $T(r,f)$ can be bounded only by an $N(r,a)$ and an $\overline{N}(r,\frac{1}{f^{(k)}-b})$. This fact is contained in the famous Hayman's inequality as follows.

THEOREM F. (1, p. 138) Let $f(z)$ be a transcendental meromorphic function, and $k$ be a positive integer. Then

$$T(r,f) < C_k\{N(r,1/f) + \overline{N}(r,\frac{1}{f^{(k)}-1})\} + O\{\log r\, T(r,f)\},$$

where $C_k$ is a constant depending only on $k$.

Zhu Jinghao proved

THEOREM 19. [27] Let $f(z)$ be as in Theorem F, and $\phi(z)$ be a meromorphic function satisfying $T(r,\phi) = o\{T(r,f)\}$. Then for any positive integer $k$, we have

$$T(r,f) < C_k[N(r,\frac{1}{f}) + \overline{N}(r,\frac{1}{f^{(k)}-\psi(z)}) + O\{\log rT(r,f)\}],$$

where $C_k$ is a constant depending only on $k$.

Gu Yongxin proved in 1979 a normality criteria corresponding to Hayman's inequality.

THEOREM G. [11] Let $\{f_\alpha\}_{\alpha\in I}$ I an index set be a family of meromorphic functions in a domain D. Let $k$ be a positive integer, and $a$, $b$ two finite complex number with $b \neq 0$. If $f_\alpha(z) \neq a$, $f_\alpha^{(k)}(z) \neq b$, then $\{f_\alpha(z)\}$ is a normal family in D.

Zhu Jinghao proved

THEOREM 20. [28] Same hypotheses as in Theorem G with $b$ being replaced by a regular function $b(z) \neq 0$. If $f(z) \neq a$, $f_\alpha^{(k)}(z) \neq b(z)$ for a positive integer $k$, then $\{f_\alpha(z)\}$ is a normal family in D.

Concerning the normal family, Ye Yasheng [29] obtained

THEOREM 21. Let $\{f_\alpha(z)\}$ be as in Theorem G, and the multiplicity of zeroes of $f(z)$ be greater than $k$. If $\{f_\alpha^{(k)}(z)\}$ is normal in D, so is $\{f_\alpha(z)\}$.

THEOREM 22. Let $\{f_\alpha(z)\}$ be a normal family of regular functions in D, and $a_{IJ}(z)$ be regular in D. Denote

$$P_{mk}(f) = \sum_{\substack{0\le i_1<i_2<\cdots<i_k<k \\ 1\le\Sigma j_i\le m}} a_{IJ}(z)\{f^{(i_1)}(z)\}^{j_1}\cdots\{f^{(i_k)}(z)\}^{j_k},$$

$I = (i_1,i_2,\ldots,i_k)$, $J = (j_1,j_2,\ldots,j_k)$. Also, let $b(\neq 0)$ be a complex number, $n \ge m+1$. If $f_n(z) \neq 0$, $P_{mk}(f_n) - f_n^m \neq b$, then $\{f_\alpha(z)\}$ is normal in D.

## IV. NEW SINGULAR DIRECTIONS

### a) Hayman direction

Based on Hayman's inequality, Yang Le defined the notion of Hayman direction for a meromorphic function $f(z)$, and discussed its existence.

Suppose that $f(z)$ is a meromorphic function of order $\rho$. A direction $H$: $\arg z = \theta$, is called a Hayman direction of $f(z)$, if for given $\varepsilon > 0$ and complex number $a$ and $b$ with $b \neq 0$,

$$\overline{\lim_{r\to\infty}} \frac{\log\{n(r,\theta,\varepsilon,f=a)+n(r,\theta,\varepsilon,f^{(k)}=b)\}}{\log r} = \rho, \qquad 0<\rho<\infty \tag{5}$$

or

$$\lim_{r\to\infty} \log\{n(r,\theta,\varepsilon,f=a)+n(r,\theta,\varepsilon,f^{(k)}=b)\} = \infty, \qquad \rho=0, \tag{6}$$

where $k$ is a positive integer.

Gu Yongxin [12], Zhang Qingde and Yang Le [13] proved that for entire or meromorphic function, of order $\rho$, $0 < \rho < \infty$ respectively, there exists at least one Hayman direction.

Ye Yasheng [36] proved

THEOREM 23. For a transcendental entire function of order zero, there exists at least one Hayman direction.

Jing Lu dealt with the existence of such direction when $f^{(k)}(z)$ is substituted by $g(z) = f^{(k)}(z) + \sum_{j=0}^{k-1} a_j(z)f^{(j)}(z)$ in (5). He proved.

THEOREM 24. [31] Suppose that $f(z)$ is an entire function of order $\rho$ $(0<\rho<\infty)$. Let $a_j(z)$, $j = 0,1,2,\dots,k-1$, be entire functions order $\rho_j$. Let $h_1(z),\dots,h_k(z)$ be an arbitrary system of basic solution of the equation

$$R(w) = w^{(k)}(z) + \sum_{j=0}^{k=1} a_j(z)w^{(j)}(z) = 0,$$

and $h_l(z)$, $l = 1,2,\dots,k$, be entire of order $\rho_l^*$, respectively. If

$$\max(\rho_0,\rho_1,\dots,\rho_{k-1};\rho_1^*,\rho^*,\dots,\rho_k^*) < \rho,$$

then there exists at least one direction $H$: $\arg z = \theta$, such that for given $\varepsilon < 0$ and two complex numbers $\alpha,\beta$ with $\alpha a_0(z) \not\equiv \beta$,

$$\overline{\lim_{r\to\infty}} \frac{\log\{n(r,\theta,\varepsilon,f=\alpha)+n(r,\theta,\varepsilon,g=\beta)\}}{\log r} = \rho,$$

where

$$g(z) = f^{(k)}(z) + \sum_{j=0}^{k-1} a_j(z)f^{(j)}(z).$$

b) Borel direction

Let $f(z)$ be a meromorphic function of order $\rho$ $(0<\rho<\infty)$. Denote by $S(\rho)$ the union of all meromorphic functions with orders less than $\rho$ and the value $\infty$. A direction $B$: $\arg z = \theta$ is called the Borel direction of $f(z)$

with respect to $S(\rho)$, if for given $\varepsilon > 0$ and $\rho_j \in S(\rho)$, $j = 1,2,3$,

$$\overline{\lim_{r\to\infty}} \frac{\log\{n(r,\theta,\varepsilon,f=\phi_1) + n(r,\theta,\varepsilon,f=\phi_2) + n(r,\theta,\varepsilon,f=\phi_2)\}}{\log r} = \rho.$$

Biernacki [14, p. 153] proved that for a meromorphic function of order $\rho$ $(0<\rho<\infty)$, there exists at least one Borel direction with respect to $S(\rho)$, which coincides with the Borel direction of $f(z)$.

Pang Xuecheng proved

THEOREM 25. [32] Let $f(z)$ be a meromorphic function of order $\rho$ $(0<\rho<\infty)$, and $\rho(r)$ be its proximate order, $U(r) = r^{\rho(r)}$. Denote $S^*(\rho(r))$ as the set consisting of all meromorphic functions $\phi(z)$ satisfying $T(r,\phi) = o\{U(r)\}$ and the value $\infty$. Then there exists at least one Borel direction of $f(z)$ with respect to $S^*(\rho(r))$.

Since $S(\rho) \subset S^*(\rho(r))$, Theorem 25 improves the result of Biernacki. Also, there is an example to show that Borel direction is not necessarily the Borel direction with respect to $S^*(\rho(r))$.

c) Nevanlinna direction

Lu Yinian and Zhang Guanghou defined the notion of Nevanlinna direction in 1982, and proved

THEOREM H. [15] Let $f(z)$ be a meromorphic function satisfying

$$\overline{\lim_{r\to\infty}} \frac{T(r)}{(\log r)^2} = \infty; \quad \underline{\lim_{r\to\infty}} \frac{\log T(r)}{\log r} = \mu < \infty,$$

then there exists at least one Nevanlinna direction of $f(z)$.

Ling Qun [26] proved

THEOREM 26. Let $f(z)$ be a meromorphic function in an angular domain $\Omega(\theta_1,\theta_2) = \{z;\ \theta_1 \le \arg z \le \theta_2\}$. If

$$\overline{\lim_{r\to\infty}} \frac{T(r,\Omega(\theta_1,\theta_2))}{(\log r)^2} = \infty,$$

then either one of Li; $\arg z = \theta_i$, $i = 1,2$, is a Julia direction, or there exists at least one Nevanlinna direction.

V. FACTORIZATION THEORY

A meromorphic function $F(z)$ is said to be prime (psuedoprime) if every factorization of the form $F = f(g)$ implies that either $f$ is linear or $g$

is linear (either $f$ is rational or $g$ is a polynomial). A transcendental meromorphic function $F(z)$ is said to be left prime, if every factorization of the above form implies that $f$ is fractional linear wherever $g$ is transcendental.

a) It has been a research problem to find the classes of prime or pseudo-prime entire or meromorphic functions.

Prokopovich [16] proved

THEOREM I. Let $P_j$, $Q_j$ be polynomials, $\deg P_j = n_j$, $0 \leq n_1 < n_2 < \ldots < n_m$, $Q_j \not\equiv 0$ $(j = 1,2,\ldots,m)$. Put

$$F(z) = \sum_{j=1}^{m} Q_j e^{P_j(z)}, \quad F(z) \not\equiv C + Q_2(z)e^{P_2(z)}.$$

Then $F(z)$ is prime, unless all $P_j$ and $Q_j$ have a common right factor, that is, there exists a non-linear polynomial $g(z)$ such that

$$P_j = p_j(g), \quad Q_j = q_j(g), \quad j = 1,2,\ldots,m,$$

where $p_j$ and $q_j$ are polynomials.

Song Guodong generalized Theorem I by given

THEOREM 27. [34] The conclusion of Theorem I remains valid of all $Q_j$'s in the theorem are rational functions.

Steinmetz proved

THEOREM J. [17] Let $h(z)$ be a transcendental meromorphic function satisfying a linear differential equation

$$W^{(n)} + a_{n-1}(z)W^{(n-1)} + \ldots + a_0(z)W + a(z) = 0, \tag{7}$$

where $a(z)$, $a_0(z),\ldots,a_{n-1}(z)$ are rational functions. Then $h(z)$ is pseudo-prime.

Song Grodong and Yang Chungchun [35] proved

THEOREM 28. Let $D$ denote the class of meromorphic functions which satisfy a linear differential equation of the form (7). Put

$$F(z) = \sum_{j=1}^{m} Q_j(z)\phi_j(z),$$

where $\phi_j(z) \in D$, and $Q_j(z)$ is rational, $j = 1,2,\ldots,m$. Then the function $F$ and all its derivatives $F^{(n)}$ $(n = 1,2,\ldots)$ are pseudo-prime.

THEOREM 29. Suppose that in addition to the hypotheses of Theorem 28 the function $\phi_j$ satisfies

$$T(r,\phi_j) = o\{T(r,F)\} \quad \text{as} \quad r \to \infty, \quad j = 1,\ldots,m-1 \qquad (m \geq 2)$$

and

$$N(r,\phi_m) + N(r, \frac{1}{\phi_m}) = o\{T(r,F)\} \quad \text{as} \quad r \to \infty,$$

where $F(z)$ is of the form

$$F(z) \equiv \sum_{j=1}^{m} Q_j(z)\phi_j(z), \quad F(z) \not\equiv Q(z) + Q_2(z)\phi_2(z)$$

with $Q_j$ being non-zero rational functions and $Q$ being any polynomial. Then the function $F$ and all its derivatives $F^{(n)}$ $(n = 1,2,\ldots)$ are left prime.

b) Ozawa's conjecture

M. Ozawa posed a conjecture in 1982: Let $F(z)$ be an entire function for which there exist polynomials $P_m(z)$ of degree $m$ and entire functions $f_m(z)$ so that

$$F(z) = P_m(f_m(z)) \tag{8}$$

for $m = n_j$, $j = 1,2,\ldots,$ and $m = q$ with $(q,n_j) = 1$, $n_j | n_{j+1}$, $j = 1,2,\ldots$ . Then $F(z)$ is either of the form

$$F(z) = ae^{H(z)} + b$$

or

$$F(z) = a \cos \sqrt{H(z)} + b,$$

where $a,b$ are constants, and $H(z)$ is an entire function.

W. Fuchs and Song Guodong [36] showed by giving an example that the Ozawa's conjecture is not valid in the general case, but they proved

THEOREM 30. If equation (8) holds for $m = q$ and $m = n_j$, where $2 \leq n_1 < q$, $(n_j,q) = 1$, $n_{j+1} \leq n_j q$, $j = 1,2,3,\ldots,$ . Then the conclusion of Ozawa's conjecture is valid.

Song Guodong and Huang Jue [37] proved

THEOREM 31. If equation (8) holds for $m = n_j$ with $(n_j,n_k) = 1$ $(j \neq k)$. Then the conclusion of Ozawa's conjecture is valid.

VI. OTHER TOPICS

a) Milloux's inequality

Milloux [6] proved

THEOREM K. Let $f(z)$ be a meromorphic function. Put

$$\phi(z) = \sum_{j=0}^{k} a_j(z) f^{(j)}(z), \qquad (k \geq 1),$$

where $a_j(z)$, $j = 0,1,\ldots,k$, are entire functions satisfying

$$a_0(z) - 1 \not\equiv 0, \quad a_1(z) \not\equiv 0, \quad T(r,a_j) = 0\{T(r,f)\}.$$

Then

$$T(r,f) < (2k+1)N(r,1/f) + (k+1)N(r,1/f-1) + N(r,1/\phi-1) + S(r,f).$$

Huang Jue and Song Guodong [38] proved

THEOREM 32. Let $f(z)$, $\phi(z)$ and $a_j(z)$ be as in Theorem J except that $a_j(z)$ may be meromorphic. Then

$$T(r,f) < (k^2+k+1)\overline{N}(r,1/f) + (k+2)\overline{N}(r,1/f-1) + \overline{N}(r,1/\phi-1) + S(r,f).$$

b) Mean value of entire functions

Shah and Silverman [7] proved

THEOREM L. Let $f(z)$ be an entire function of finite lower order. Denote

$$M_0(r) = e^{T(r,f)}, \quad M_s(r) = \left\{\frac{1}{2\pi}\int_0^{2\pi} |f(re^{i\theta})|^s d\theta\right\}^{\frac{1}{s}}, \quad 0 < s < \infty.$$

If $\sum_{a\neq\infty} \delta(a,f) = 1$, then for any $\varepsilon > 0$

$$\lim_{r\to\infty} \frac{\{M_0(r)\}^{\pi-\varepsilon}}{M_s(r)} = 0. \tag{9}$$

Jing Lu obtained

THEOREM 33. [39] Let $f(z)$ be as in Theorem K, and $a(z)$ $(\not\equiv \infty)$ be any meromorphic function satisfying $T(r,a(z)) = \{T(r,f)\}$. If $\Sigma\delta(a(z),F) = 1$, then (9) remains valid.

REFERENCES

1. Yang Le, Value distribution theory and its new research (Chinese), 1982, Academic press.

2. Yang Le and Zhang Guanghou, Sci. Sinica, 18 (1975), 23-37.

3. Yang Le and Zhang Guanghou, Acta Math. Sinica (Chinese), 18 (1975), 35-53.

4. F. Nevanlinna, Septieme Congress Math. Scand. Oslo. (1930) 81-83.

5. Yang Lo, Sci. Sinica, 4 (1981), 394-404.

6. H. Milloux, J. Math, Pares et appliq., 19 (1940), 197-210.

7. S. M. Shah and H. Silverman, Math, Ann. 220 (1972), 185-192.

8. Zhuang Qitai, Sci. Sinica, 13 (1964), 887-895.

9. S. M. Shah and H. Silverman, Math. Ann. 220 (1976), 185-192.

10. A. Weitsman, Acta Math., 128 (1972), 41-52.

11. Gu Yongxin, Sci. Sinica, Special edition for Mathematics (I), 1979, 267-274.

12. Gu Yongxin, Acta Math. Sinica (Chinese), 25 (1982), 28-48.

13. Zhang Qingde and Yang Lo, Sci. Sinica, 11 (1983).

14. Zhuang Qitai, Singular direction of meromorphic functions (Chinese), 1982, Academic press.

15. Lu Yinian and Zhang Guanghou, Sci. Sinica, 3 (1983), 215-224.

16. G. S. Prokopovich, Ukr. Math. Z. 26, 2 (1974), 188-195.

17. N. Steinmetz, Math. Z. 170 (1980), 169-180.

18. Zhang Qingde and Huang Jue, Jour. East China Normal Univ., 2 (1980), 36-49.

19. Zhang Qingde and Pang Xuecheng; Relation between the number of proximate deficient functions and that of common Borel directions for a meromorphic function, to appear.

20. Pang Xuecheng and Ru Ming, Total number of deficient entire functions of an entire function, to appear.

21. Pang Xuecheng and Ru Ming, Relation between the number of proximate deficient functions and that of common Borel directions for a certain class of meromorphic functions, to appear.

22. Li Qingzhong and Ye Yasheng, Sum of deficiency of deficient functions and F. Nevanlinna's conjecture, to appear.

23. Jing Lu and Dai Chongji, On Nevanlinna's conjectures concerning deficient functions, to appear.

24. Jing Lu and Dai Chongji, On Nevanlinna's conjectures concerning deficient functions, II, to appear.

25. Ling Qun and Dai Chongji, On S. M. Shah conjecture, to appear.

26. Ling Qun, Nevanlinna directions in an angular domain, to appear.

27. Zhu Jinghao, General form of Hayman's inequality, to appear.

28. Zhu Jinghao, On normality critirion of meromorphic functions, to appear.

29. Ye Yasheng, Some results concerning normal family, to appear.

30. Ye Yasheng, On Hayman's direction of meromorphic functions, to appear.

31. Jing Lu, On angular distribution of entire functions, to appear.

32. Pang Xuecheng, On singular direction of meromorphic functions, to appear.

33. Ling Qun, On convergence of $\Sigma\delta^{1/3}(a(z),f)$ for deficient functions, to appear.

34. Song Guodong, On primality of the combination of exponential functions, to appear.

35. Song Guodong and Chung-Chun Yang, Contemporary Mathematics, American Math. Soc. (Edited by Chung-Chun Yang), Vol. 25 (1983), 155-112.

36. W. H. J. Fuchs and G. D. Song, On a conjecture by M. Ozawa concerning factorization of entire functions, to appear.

37. Song Guodong and Huang Jue, On Ozawa's conjecture concerning the functional equation $F(z) = P_k(f_k)$.

38. Huang Jue and Song Guodong, Jour. East China Normal Univ., 3 (1984), 1-5.

39. Jing Lu, Entire functions which have the maximum sum of deficiency of deficient functions and their mean values, to appear.

40. Dai Chongji, Jour. East China Normal Univ., 1 (1979).

41. Dai Chongji and Ji Shanyu, Jour. East China Normal Univ., 2 (1980).

42. Zhang Guanghou, Acta Math. Sinica (Chinese), 20 (1977), 73-98.

43. Dai Chongji and Ji Shanyu, Jour. East China Normal Univ., 1 (1981).

44. Ji Shanyu, Jour. East China Normal Univ., 2 (1981).

45. Zhang Qingde, Jour. East China Normal Univ., 4 (1983).

46. Zhang Qingde, Jour. East China Normal Univ., 1 (1984).

Department of Mathematics
East China Normal University
Shanghai

Contemporary Mathematics
Volume 48, 1985

# UNIVALENT FUNCTIONS

Shu-qin Liu and Chi-tai Chuang

In China the research on univalent functions was initiated by professor Jian-kung Chen. As early as the year of 1936, he published two papers [1,2] on univalent functions and his work was succeeded by some of his students. In past years, many Chinese mathematicians have worked in this field and obtained many significant results of which the present article gives a brief account.

Contents

I. The classes $S$ and $\Sigma$.
II. Generalizations of the classes $S^*$ and $S^c$.
III. Symmetric univalent functions.
IV. Some special classes.
V. Meromorphic univalent functions.
VI. Typically real functions.
VII. Results on subordination.
VIII. Univalent functions in multiply connected domains.

## I. THE CLASSES $S$ AND $\Sigma$

### 1.1. General inequalities

Let $S$ be the class of functions

$$f(z) = z + \sum_{n=2}^{\infty} a_n z^n$$

holomorphic and univalent in $D : |z| < 1$ and let $\Sigma$ be the class of functions

$$F(\zeta) = \zeta + \sum_{n=0}^{\infty} \alpha_n \zeta^{-n}$$

meromorphic and univalent in $\Delta : |\zeta| > 1$.

If $f(z) \in S$ then $F(\zeta) = 1/f(\zeta^{-1}) \in \Sigma$.

Throughout this article we shall denote by $D$ and $\Delta$ the unit disk $|z| < 1$ and the domain $|\zeta| < 1$ respectively.

Let $F(\zeta) \in \Sigma$ and define

$$Q_F(\zeta,\zeta') = \frac{F(\zeta')-F(\zeta)}{\zeta'-\zeta} \ (\zeta\neq\zeta'), \quad Q_F(\zeta,\zeta) = F'(\zeta).$$

Grunsky and Golusin proved

THEOREM 1.1. If $F(\zeta) \in \Sigma$ and $|\zeta_\nu| > 1$ $(\nu=1,2,\dots,n)$, then

$$\left|\sum_{\nu,\mu=1}^{n} \alpha_\nu\alpha_\mu \log Q_F(\zeta_\nu,\zeta_\mu)\right| \le -\sum_{\nu,\mu=1}^{n} \alpha_\nu\bar{\alpha}_\mu \log\left(1-\frac{1}{\zeta_\nu\bar{\zeta}_\mu}\right). \tag{1.1}$$

Golusin extended (1.1) to

THEOREM 1.2. Let $a_{\nu\mu}$ be the coefficients of the positive definite quadratic form $\sum_{\nu,\mu=1}^{n} a_{\nu\mu}x_\nu x_\mu$, then

$$\prod_{\nu,\mu=1}^{n}\left|1-\frac{1}{\zeta_\nu\bar{\zeta}_\mu}\right|^{a_{\nu\mu}} \le \prod_{\nu,\mu=1}^{n}\left|Q_F(\zeta_\nu,\zeta_\mu)\right|^{a_{\nu\mu}} \le \prod_{\nu,\mu=1}^{n}\left|1-\frac{1}{\zeta_\nu\bar{\zeta}_\mu}\right|^{-a_{\nu\mu}}. \tag{1.2}$$

Shah [9] unified (1.1) and (1.2) to

$$\left|\sum_{\nu,\mu=1}^{n} a_{\nu\mu}\alpha_\nu\alpha_\mu \log Q_F(\zeta_\nu,\zeta_\mu)\right| \le \sum_{\nu,\mu=1}^{n} a_{\nu\mu}\alpha_\nu\bar{\alpha}_\mu \log\left(1-\frac{1}{\zeta_\nu\bar{\zeta}_\mu}\right)^{-1}. \tag{1.3}$$

Shah [10] also gave the extension:

THEOREM 1.3. If $F_k(\zeta) \in \Sigma$ $(k=1,2,\dots,m)$ and $\sum_{\nu,\mu=1}^{n} a_{\nu\mu}x_\nu x_\mu \ge 0$, then

$$\left|\sum_{\nu,\mu=1}^{n} a_{\nu\mu}\alpha_\nu\alpha_\mu \prod_{k=1}^{m} \log Q_{F_k}(\zeta_\nu,\zeta_\mu)\right| \le \sum_{\nu,\mu=1}^{n} \alpha_\nu\alpha_\mu\left(\log\frac{1}{1-\frac{1}{\zeta_\nu\bar{\zeta}_\mu}}\right)^{m}. \tag{1.4}$$

Golusin [11] proved that

$$\left|\log Q_F(\zeta,\zeta')\right| \le \log\sqrt{\frac{1}{1-|\zeta|^{-2}}\,\frac{1}{1-|\zeta'|^{-2}}}\ . \tag{1.5}$$

Shah [12] improved (1.5) to

$$\left|\log Q_F(\zeta,\zeta')\right| \le \sqrt{\log\frac{1}{1-|\zeta|^{-2}}\,\log\frac{1}{1-|\zeta'|^{-2}}}\ . \tag{1.6}$$

FitzGerald [13] obtained the "exponentiated" form of (1.1). His theorems may be stated as follows:

THEOREM 1.4. If $F(\zeta) \in \Sigma$, then

$$\left| \sum_{\nu,\mu=1}^{n} \alpha_\nu \alpha_\mu Q_F(\zeta_\nu, \zeta_\mu) \right| \le \sum_{\nu,\mu=1}^{n} \alpha_\nu \overline{\alpha}_\mu \frac{\zeta_\nu \overline{\zeta}_\mu}{\zeta_\nu \overline{\zeta}_\mu - 1}. \tag{1.7}$$

THEOREM 1.5. Let $f(z) \in S$, $\{z_\nu\}_{\nu=1}^{n} \in D$. $U_{\nu,\mu} = (f(z_\nu)-f(z_\mu)) \cdot (z_\nu - z_\mu)^{-1}$, $V_\nu = f(z_\nu) z_\nu^{-1}$, $W_{\nu,\mu} = (1 - z_\nu \overline{z}_\mu)^{-1}$, then for all complex numbers $\beta_\nu (\nu=1,2,\ldots,m)$, we have

$$\left| \sum_{\nu=1}^{m} \beta_\nu |v_\nu|^i \right|^2 \le \sum_{\nu,\mu=1}^{m} \beta_\nu \overline{\beta}_\mu |U_{\nu,\mu} W_{\nu,\mu}|^i \quad (i=1,2). \tag{1.8}$$

When $z_\nu = z_\mu$, $U_{\nu,\mu}$ is defined to be $f'(z_\nu)$.

Kung [14] improved (1.8) to

1.6. Let

$$\Omega(z_\nu) = -f(z_\nu)^{-1} + z_\nu^{-1} - a_2 + \overline{z}_\nu,$$

$$\Omega'(z_\nu) = -f(z_\nu)^{-1} + z_\nu^{-1} - a_2 - \overline{z}_\nu, \tag{1.9}$$

then with the hypothesis of the above theorem, we have

$$\sum_{\nu,\mu=1}^{m} \beta_\nu \overline{\beta}_\mu |U_{\nu,\mu} W_{\nu,\mu}|^\ell \ge \sum_{\nu,\mu=1}^{m} \beta_\nu \overline{\beta}_\mu |V_\nu V_\mu|^\ell \exp \tfrac{\ell}{2} \Omega(z_\nu) \overline{\Omega}(z_\mu) \tag{1.10}$$

$$\sum_{\nu,\mu=1}^{m} \beta_\nu \overline{\beta}_\mu |U_{\nu,\mu}^{-1} W_{\nu,\mu}|^\ell \ge \sum_{\nu,\mu=1}^{m} \beta_\nu \overline{\beta}_\mu |V_\nu^{-1} V_\mu^{-1}|^\ell \exp \left\{ \tfrac{\ell}{2} \Omega'(z_\nu) \overline{\Omega}'(z_\mu) \right\}, \tag{1.11}$$

where $\ell \ge 0$.

Set

$$\zeta = \frac{1}{z}, \quad F(\zeta) = f\left(\frac{1}{\zeta}\right)^{-1} = \zeta + \sum_{n=0}^{\infty} A_n \zeta^{-n},$$

$$-\Omega(z) = f(z)^{-1} - z^{-1} + a_2 - \overline{z} = F(\zeta) - \zeta - A_0 - \overline{\zeta}^{-1} = \sum_{n=1}^{\infty} A_n \zeta^{-n} - \overline{\zeta}^{-1}$$

$$= G(\zeta) = g(z),$$

$$-\Omega'(z) = f(z)^{-1} - z^{-1} + a_2 + \overline{z} = F(\zeta) - \zeta - A_0 + \overline{\zeta}^{-1} = \sum_{n=1}^{\infty} A_n \zeta^{-n} + \overline{\zeta}^{-1}$$

$$= H(\zeta) = h(z).$$

Then (1.10) and (1.11) may be written as

$$\sum_{\nu,\mu=1}^{m} \beta_\nu \bar{\beta}_\mu \left| U_{\nu,\mu} W_{\nu,\mu} \right|^{\ell} \geq \sum_{p=0}^{m} \frac{1}{p!}\left(\frac{\ell}{2}\right)^p \left| \sum_{\nu=1}^{m} \beta_\nu \left| V_\nu \right|^{\ell} g^p(z_\nu) \right|^2 \tag{1.12}$$

$$\sum_{\nu,\mu=1}^{m} \beta_\nu \bar{\beta}_\mu \left| U_{\nu,\mu}^{-1} W_{\nu,\mu} \right|^{\ell} \geq \sum_{p=0}^{m} \frac{1}{p!}\left(\frac{\ell}{2}\right)^p \left| \sum_{\nu=1}^{m} \beta_\nu \left| V_\nu^{-1} \right|^{\ell} h^p(z_\nu) \right|^2 \tag{1.13}$$

respectively. In the right of (1.12), take the first terms only and let $\ell = 1,2$, we get the inequalities (1.8).

Kung [15] also improved FitzGerald inequalities from another point of view.

We point out that Hu [17] had already obtained FitzGerald inequality (1.8) in 1958, but unfortunately for some reasons, he could not be able to publish his paper at that time. Hu [17] also gave some generalizations of the inequalities (1.8), one of which may be stated as follows:

Let $f(z) \in S$,

$$\log \frac{f(z)-f(\zeta)}{z-\zeta} \frac{z\zeta}{f(z)f(\zeta)} = \sum_{n,m=1}^{\infty} d_{nm} z^n \zeta^m \quad (|z| < 1, \quad |\zeta| < 1),$$

$$g_m^{\varepsilon}(z) = -F_m\left(\frac{1}{f(z)}\right) + \frac{1}{z^m} + \varepsilon \bar{z}^m = m \sum_{n=1}^{\infty} d_{nm} z^n + \varepsilon \bar{z}^m \quad (\varepsilon = \pm 1),$$

where $F_m(t)$ is the Faber polynomial of degree $m$ produced by $t = 1/f(z)$. Let

$$f_\mu = f(z_\mu), \quad \varphi_\varepsilon(z_\mu, z_\nu) = \left| \frac{f_\mu - f_\nu}{z_\mu - z_\nu} \right|^{\varepsilon} \frac{1}{1 - z_\mu \bar{z}_\nu}.$$

THEOREM 1.7. For any complex sequence $x_\nu$ $(\nu=1,2,\ldots,n)$ if $\sum_{\nu,\mu=1}^{n} a_{\nu,\mu} x_\nu \bar{x}_\mu \geq 0$, then for any $\alpha > 0$ and any positive integer $m$, we have

$$\sum_{\nu,\mu=1}^{n} a_{\mu,\nu} x_\mu \bar{x}_\nu \left| \frac{f_\mu f_\nu}{z_\mu z_\nu} \right|^{\alpha\varepsilon} \exp\left\{ \frac{\alpha}{2} \sum_{n=1}^{m} \frac{1}{n} g_n^{\varepsilon}(z_\mu) \overline{g_n^{\varepsilon}(z_\nu)} \right\} \tag{1.14}$$

$$\sum_{\mu,\nu=1}^{n} a_{\mu,\nu} x_\mu \bar{x}_\nu \phi_\varepsilon^{\alpha}(z_\mu, z_\nu), \quad (|z_\nu| < 1, \ \nu=1,2,\ldots,n).$$

In particular, when $a_{\mu,\nu} \equiv 1$, $m = 1$, we get Kung's result, and when $a_{\mu,\nu} \equiv 1$, $m = 0$, we get FitzGerald inequality.

Hu [19] also proved the following.

THEOREM 1.8. If $f(z) \in S$, $f_2(z) = (f(z^2))^{\frac{1}{2}} = z + \sum_{n=1}^{\infty} a_{2n+1}^{(2)} z^{2n+1} \in S_2$, where $S_2$ denotes the family of odd univalent functions, then in letting

$$\left\|\frac{f(\rho)}{\rho}\right\| = 1 + \sum_{n=2}^{\infty} |a_n| \rho^{n-1}, \quad 0 < \rho < 1,$$

we have

$$\left\|\frac{f(\rho^2)}{\rho^2}\right\| \le \left(\frac{1}{1-\rho^2}\,\frac{1}{2\pi}\int_0^{2\pi}\left|\frac{f(\rho e^{i\theta})}{\rho}\right| d\theta\right)^2 \left(\sum_{k=1}^{n}\left|a_{2k+1}^{(2)}\right|^2 \frac{1}{1-\rho^2} - \frac{1}{2\pi}\int_0^{2\pi}\left|\frac{f(\rho e^{i\theta})}{\rho}\right| d\theta\right)^2 .$$

Ren [20] improved Grunsky inequalities as follows:

THEOREM 1.9. If $f(z) \in S$, $F(\zeta) \in \Sigma$, $|z|, |z'| < 1$, $|\zeta|, |\zeta'| > 1$ and let

$$\log \frac{f(z)-f(z')}{z-z'} = \sum_{k,\ell=0}^{\infty} c_{k\ell} z^k z'^{\ell}, \quad \log \frac{F(\zeta)-F(\zeta')}{\zeta-\zeta'} = \sum_{k,\ell=1}^{\infty} r_{k\ell} \zeta^{-k} \zeta'^{-\ell},$$

then for any complex numbers $\lambda_k$ $(k=1,2,\dots,n)$, $\lambda'_\ell$ $(\ell=1,2,\dots,n)$ and positive integers $m$, $n$, $n'$, we have

$$\left|\sum_{k=1}^{n}\sum_{\ell=1}^{n'} (c_{k\ell})^m \lambda_k \lambda'_\ell\right|^2 \le \sum_{k=1}^{n} \frac{|\lambda_k|^2}{k^m} \cdot \sum_{\ell=1}^{n'} \frac{|\lambda_\ell|^2}{\ell^m},$$

$$\left|\sum_{k=1}^{n}\sum_{\ell=1}^{n'} (r_{k\ell})^m \lambda_k \lambda'_\ell\right|^2 \le \sum_{k=1}^{n} \frac{|\lambda_k|^2}{k^m} \cdot \sum_{\ell=1}^{n'} \frac{|\lambda'_\ell|^2}{\ell^m}.$$

When $n' = n$, $|\lambda_k| = |\lambda'_k|$ and $\{\lambda_k\}$, $\{\lambda'_k\}$ are real, then the above inequalities can be equalities for $f(z) = z/(1\pm z)^2$. For $m = 1$, they reduce to Grunsky inequalities.

Ren [107] also obtained generalizations of the FitzGerald inequalities and proved, among others, the following theorems:

THEOREM 1.10. Let $f(z) \in S$, $\{r_{kn}\}$ the Grunsky coefficients of

$$g(\zeta) = \frac{1}{f(\frac{1}{\zeta})} = \zeta + \sum_{n=0}^{\infty} b_n \zeta^{-n},$$

$\theta_0 = \theta_0(f)$, $\alpha_f$ the Hayman direction and Hayman constant of $f(z)$ respectively and

$$\log(f(z)/z) = 2\sum_{n=1}^{\infty} r_n z^n.$$

If $\sum_{k=1}^{\infty} |\eta_k|^2/k < \infty$, then we have the sharp inequality

$$\left(\frac{1}{2}\log\frac{1}{\alpha_f} - \sum_{n=1}^{\infty} n\left|r_n - \frac{1}{n}e^{-in\theta_0}\right|^2\right)\left(\sum_{k=1}^{\infty}\frac{1}{k}|\eta_k|^2 - \sum_{n=1}^{\infty} n\left|\sum_{k=1}^{\infty}\eta_k r_{kn}\right|^2\right)$$

$$\geq \left|\sum_{k=1}^{\infty}\eta_k\left[\left(\frac{1}{k}e^{-ik\theta_0} - r_k\right) + \sum_{n=1}^{\infty} n\left(\overline{r_n - \frac{1}{n}e^{-in\theta_0}}\right) r_{kn}\right]\right|^2. \tag{1.15}$$

This theorem improves the famous Bazilevic inequality [108] and one of Goluzin. In particular, it contains [6] the inequality

$$\left(\frac{1}{2}\log\frac{1}{\alpha_f} - \sum_{n=1}^{\infty} n\left|r_n - \frac{1}{n}e^{-n\theta_0}\right|^2\right)\left(1 - \sum_{n=1}^{\infty} n|b_n|^2\right)$$

$$\geq \left|r_1 - e^{-i\theta_0} + \sum_{n=1}^{\infty} n\left(\overline{r_n - \frac{1}{n}e^{-in\theta_0}}\right) b_n\right|^2. \tag{1.16}$$

THEOREM 1.11. Let $w = F(\zeta) = \zeta + \sum_{n=1}^{\infty} \alpha_n \zeta^{-n} \in \Sigma$ and $G(w) = F^{-1}(w) = w + \sum_{n=1}^{\infty} \beta_n w^{-n}$ be the inverse function of $w = F(\zeta)$. Let

$$\log\frac{G(w)-G(w')}{w-w'} = \sum_{p,q=1}^{\infty} \lambda_{pq} w^{-p} w'^{-q}.$$

Then we have

$$\left|\sum_{m,n=1}^{\infty} \lambda_{mn} x_m x_n\right| \leq \sum_{n=1}^{\infty} \frac{1}{n}\left|x_n + \sum_{m=n+2}^{\infty} \beta_m^{(-n)} x_m\right|^2,$$

where $\beta_n^{(-k)}$ is defined by

$$[G(w)]^{-k} = w^{-k} + \sum_{n=k+2}^{\infty} \beta_n^{(-k)} w^{-n}.$$

This theorem is useful in view of the Springer conjecture. (see next paragraph).

Finally we point out that besides his works mentioned above, Hu [86, 89] also established theorems concerning functions of the form $\varphi(z) = e^{w(z)}$, $w(z) = \sum_{n=1}^{\infty} A_n z^n$, which generalize results of Milin.

1.2. Coefficients problems and covering theorems.

In the following we denote as usual by $S^*$ the class of those functions $f(z) \in S$ such that $f(z)$ maps $D$ onto a starlike domain with respect to $w = 0$, and by $S^c$ the class of those functions $f(z) \in S$ such that $f(z)$ maps $D$ onto a convex domain.

A. Coefficients problems

The Bieberbach conjecture $|a_n| \le n$ concerning the class $S$ was investigated by some Chinese mathematicians in the works [110, 32, 33, 35, 38, 39, 40, 41, 45]. However, we do not state their results here, in view of a recent work of de Branges.

If $F(\zeta) = \zeta + \sum_{n=0}^{\infty} \alpha_n \zeta^{-n} \in \Sigma$, Schiffer [46] proved that $|\alpha_2| \le \frac{2}{3}$. Garabedian and Schiffer [47] proved that $|\alpha_3| \le \frac{1}{2} + e^{-6}$. When $n \ge 4$, the problem is open.

Let the inverse function of $F(\zeta)$ be $G(w) = w + \sum_{n=1}^{\infty} \beta_n w^{-n}$. Springer [48] proved $|\beta_3| \le 1$ and conjectured

$$|\beta_{2k-1}| \le \frac{(2k-2)!}{k!(k-1)!} \qquad k = 1,2,\ldots, \tag{1.17}$$

equality is valid only for the inverse function of $w = \zeta + \frac{\eta}{\zeta}$, $|\eta| = 1$. Kubota [49], using Grunsky inequality, proved (1.17) for $k = 3,4,5$. Ren [50, 51] proved (1.17) for $k = 6,7,8$. For $k = 9$, (1.17) was proved by Yao [52].

Goluzin [82] showed that for functions $f(z) = z + \sum_{n=2}^{\infty} a_n z^n \in S$, we have

$$\big||a_{n+1}| - |a_n|\big| \le An^{\frac{1}{4}} \log n, \quad n = 2,3,\ldots, \tag{1.18}$$

and for odd functions $f(z) = \sum_{n=1}^{\infty} a'_{2n-1} z^{2n-1} \in S$, we have

$$||a'_{n+1}| - |a'_{n-1}|| \le A'n^{-\frac{1}{4}} \log n, \quad n = 2,4,\ldots, \tag{1.19}$$

where $A$ and $A'$ are constants. He also proved for functions $h(z) = z + \sum_{n=2}^{\infty} c_n z^n \in S^*$, we have

$$||c_{n+1}| - |c_n|| \le K \quad (K < 100). \tag{1.20}$$

Zhang [71] improved (1.20) to $K < 14$. Hayman improved the right side of (1.18) to an absolute constant $B$. For this, Hu [84] also gave a simple proof. Grinspan showed that $B < 3.61$. For functions $h(z) \in S^*$, Leungyuk [91] proved the inequalities

$$\begin{aligned} &|n|c_n| - m|c_m|| \le n^2 - m^2, \quad && n > m > 1, \\ &||c_{n+1}| - |c_n|| \le 1, && n = 2,3,\ldots . \end{aligned} \tag{1.21}$$

The inequalities (1.21) was also obtained by Hu.

Let $f(z) \in S$ and, $p$ being a real number, let

$$\left\{\frac{f(z)}{z}\right\}^p = 1 + \sum_{n=1}^{\infty} D_n(p) z^n.$$

Hu [88] proved the following inequalities:

$$||D_{n+1}(p)| - |D_n(p)|| \le An^{\frac{1}{2}[\tau(p)-1]} \log^{\frac{3}{2}} n, \quad n = 2,3,\ldots, \tag{1.22}$$

where $A$ is an absolute constant, $\frac{1}{4} < p < 1$ and

$$\begin{aligned} &T(p) = \frac{4p-1}{2p+t(p)}\, t(p), \quad t(p) = (\sqrt{2}\,p - 1)^2. \\ &\sum_{k=1}^{n} \Big||D_{k+1}(p)| - |D_k(p)|\Big| k^{1-p} \le An, \quad n = 2,3,\ldots, \end{aligned} \tag{1.23}$$

where equality holds for $f(z) = z/(1-z)^2$. When $p = \frac{1}{2}$, we get a result of Milin.

In particular, when $f(z) \in S^*$, we have

$$||D_{n+1}(p)| - |D_n(p)|| \le An^{p-1}, \quad n = 2,3,\ldots, \tag{1.24}$$

where the exponent $p - 1$ is the best possible, and the extremal function is

$z/(1-z)^2$. Hu [33, 85] also proved analogous inequalities concerning the coefficients $D_n(p)$ for functions $f(z) \in S$ such that

$$\lim_{\rho\to 1} \frac{(1-\rho)^2}{\rho} \max_{|z|=\rho} |f(z)| = \alpha_f \geq \alpha > 0.$$

B. Covering theorems

Consider a function $w = f(z) \in S$ which maps $D$ onto a domain $G_f$ of the w-plane. Let $w_\nu$ $(\nu=1,2,\ldots,n)$ be $n$ points of the w-plane, not belonging to $G_f$ and satisfying $\arg \frac{w_{\nu+1}}{w_\nu} = \frac{2\pi}{n}$ $(\nu=1,2,\ldots,n;\ w_{n+1} = w_1)$. Denote the lower bound of $\max(|w_1|, |w_2|,\ldots,|w_n|)$ by $A_f^{(n)}$. Szegö [111] proposed the problem to find $\min_{f\in S} A_f^{(n)}$. For $n = 1$, this problem is that concerning the Koebe constant. Lavrentieff and Chepeleff proved that

$$\min_{f\in S} A_f^{(n)} = \sqrt[n]{\frac{1}{4}},$$

and this minimum is attained only by

$$f(z) = z/(1-\varepsilon z^n)^{\frac{2}{n}}, \quad |\varepsilon| = 1. \tag{1.25}$$

For functions of the classes $S^*$ and $S^c$, Shah [113] proved the following theorems:

THEOREM 1.12. Let $f(z) \in S^*$, $n$ a positive integer $w_\nu(\nu=1,2,\ldots,m;\ m \geq n)$ be points of the w-plane not belonging to $G_f$ and satisfying $0 \leq \arg\frac{w_{\nu+1}}{w_\nu} \leq \frac{2\pi}{n}$ $(\nu=1,2,\ldots,m;\ w_{m+1} = w_1)$ and $\delta_n(f)$ the lower bound of $\max_{1\leq\nu\leq m} |w_\nu|$ when the points $w_\nu$ $(\nu=1,2,\ldots,m)$ vary. Then

$$\min_{f\in S^*} \delta_n(f) = \sqrt[n]{\frac{1}{4}},$$

and this minimum is attained only by the function (1.25).

Theorem 1.13. Let $f(z) \in S^c$, $n$ a positive integer, $w_\nu(\nu=1,2,\ldots,m;$ $m \geq n)$ points on the boundary of $G_f$ such that $0 \leq \arg\frac{w_{\nu+1}}{w_\nu} \leq \frac{2\pi}{n}$ $(\nu=1,2,\ldots,m;\ w_{n+1} = w_1)$ and $\delta_n(f)$ be the lower bound of $\max_{1\leq\nu\leq m} |w_\nu|$ when the points $w_\nu$ $(\nu=1,2,\ldots,m)$ vary on the boundary of $G_f$. Then

$$\min_{f\in S^c} \delta_n(f) = \int_0^1 \frac{dt}{(1+t^n)^{2/n}} = \frac{\Gamma^2(\frac{1}{n})}{2n\Gamma(\frac{2}{n})},$$

and this minimum is attained only by

$$f(z) = \int_0^z \frac{dt}{(1+\varepsilon t^n)^{2/n}}, \quad |\varepsilon| = 1.$$

Besides these two theorems, Shah also proved other analogous covering theorems for functions $f(z) \in S^c$, in which the domain $G_f$ is assumed to satisfy certain conditions.

1.3. Other results.

Among the functions $f(z) = z + \sum_{n=2}^{\infty} a_n z^n \in S$, those for which the second coefficient $a_2$ has a fixed modulus $|a_2|$ form a subclass of $S$ denoted by $S(|a_2|)$. Lebedev and Milin proved

THEOREM 1.14. Let $f(z) \in S(|a_2|)$, $A|a_2| = \frac{1}{2}(2 - |a_2|)^2$, $|z| = r < 1$, then

$$|f(z)| \le \frac{r}{(1-r)^2} e^{-rA|a_2|}, \qquad |f'(z)| \le \frac{1+r}{(1-r)^3} e^{-r^2 A|a_2|}.$$

This theorem was improved by Kung [22] as follows:

THEOREM 1.15. Under the same hypothesis of the above theorem, one has

$$|f(z)| \le \frac{r}{(1-r)^2} e^{-\frac{(1+r)^2}{4} A|a_2|}, \qquad |f'(z)| \le \frac{1+r}{(1-r)^3} e^{-\frac{1+r^2}{2} A|a_2| r}.$$

He also proved when $f(z) \in S(|a_2|)$ satisfies

$$|f(z)| \le \frac{r}{(1-r)^2} e^{-A|a_2|B(r)},$$

then one has

$$|f'(z)| \le \frac{1+r}{(1-r)^3} e^{-\frac{2(1+r^2)r}{(1+r)^2} B(r)A|a_2|},$$

where

$$B(r) = 2r\left[2 + \frac{1}{2}\left(\frac{(1-r)^2}{r}\right)^2 \log\frac{1+r}{1-r} - \frac{(1-r)^2}{r}^2\right]^{-1}.$$

It is well known that if a function $f(z) = z + \sum_{n=2}^{\infty} a_n z^n$ holomorphic in D is such that $\sum_{n=2}^{\infty} n|a_n| \leq 1$, then $f(z)$ is univalent in D. Thus the size of the modulus of $a_n$ has influence on the univalence of the function $f(z)$. It is interesting that this is also true for the argument of $a_n$ as shown by the following theorem of $Q_{in}$ [23]:

THEOREM 1.16. Let $f(z) = z + \sum_{n=2}^{\infty} a_n z^n$ be holomorphic in D. If for a positive integer N, we have $N|a_N| > 1$, then one can select real numbers $\alpha$ and $\beta$ such that

$$z + e^{id}\sum_{n=2}^{\infty} a_n z^n \overline{\in} S$$

and

$$z + \sum_{n=2}^{N-1} a_n z^n + e^{i\beta} a_N z^N + \sum_{n=N+1}^{\infty} a_n z^n \overline{\in} S.$$

Li [24] gave extensions of the above theorem. His main results are:

If for m positive integers $n_t$ $(t=1,2,\ldots,m)$ we have $\sum_{t=1}^{m} n_t|a_{n_t}| > 1$, then there exist real numbers $\varphi_t$ $(t=1,2,\ldots,m)$ such that for all $\alpha$ in a certain interval $(a,b)$ containing the origin, we have

$$z + e^{i\alpha}\left\{\sum_{t=1}^{m} e^{i\varphi_t} a_{n_t} z^{n_t} + \sum_{n=2}^{\infty}{}' a_n z^n\right\} \overline{\in} S,$$

where in $\sum_{n=2}^{\infty}{}'$, $n \neq n_1, n_2, \ldots, n_m$.

If $\sum_{n=2}^{\infty} n|a_n| > 1$, then there exists a sequence of real numbers $\varphi_n (n=2,3,\ldots)$ such that for all $\alpha$ in a certain interval $(a,b)$ containing the origin, we have

$$z + e^{i\alpha}\sum_{n=2}^{\infty} e^{i\varphi_n} a_n z^n \overline{\in} S.$$

Moreover, for any assigned $n_1$, $n_2$ we can make $\varphi_{n_1} = \varphi_{n_2}$.

In [25], Li pointed out that the result in [24] may be extended to the case of functions $F(\zeta) \in \Sigma$. In 1978, Silverman, Silvia and Talage [26] obtained similar results for the class of "locally univalent functions" in $D$. However, their results can also be proved by the method used in [24].

## II. GENERALIZATIONS OF THE CLASSES $S^*$ AND $S^c$

It is well known that the necessary and sufficient condition for $f(z) \in S^*$ is

$$\operatorname{Re}\left\{\frac{zf'(z)}{f(z)}\right\} \geq 0$$

in $D$, and that for $f(z) \in S^c$ is

$$1 + \operatorname{Re}\left\{\frac{zf''(z)}{f'(z)}\right\} \geq 0$$

in $D$.

Consider a number $\rho$ $(0 \leq \rho < 1)$ and let $S^*_\rho$ be the class of those functions $f(z) \in S^*$ such that

$$\operatorname{Re}\left\{\frac{zf'(z)}{f(z)}\right\} > \rho$$

in $D$. Wu [72] proved.

THEOREM 2.1. If $f(z) \in S^*_{1/2}$, then $f(z)z^{-1} < (1-z)^{-1}$, where the symbol $<$ denotes subordination, and there exists an increasing function $\alpha(\theta)$ such that

$$f(z) = \frac{1}{2\pi}\int_0^{2\pi} \frac{z}{1-ze^{-i\theta}}\, d\alpha(\theta), \qquad \alpha(\theta+0) = \alpha(\theta).$$

Then in basing upon this theorem, he proved

2.2. If $f(z) \in S^*_{1/2}$, then for $|z| = r < 1$ we have the precise estimates:

$$\operatorname{Re}\left\{\frac{f(z)}{z}\right\} \geq \frac{1}{1+r}, \quad \frac{1}{1+r} \leq \left|\frac{f(z)}{z}\right| \leq \frac{1}{1-r},$$

$$\left|\arg\frac{f(z)}{z}\right| \leq \arcsin r,$$

$$\frac{1}{1+r} \leq \left|\frac{zf'(z)}{f(z)}\right| \leq \frac{1}{1-r}, \quad \left|\arg\frac{zf'(z)}{f(z)}\right| \leq \arcsin r,$$

$$\frac{1}{(1+r)^2} \leq |f'(z)| \leq \frac{1}{(1-r)^2}, \quad |\arg f'(z)| \leq 2 \text{ arc sinr}.$$

For these estimates the extremal function is

$$f(z) = z(1-\eta z)^{-1}, \quad |\eta| = 1.$$

Wu also proved the following theorems.

THEOREM 2.3. If $f(z) \in S^*_\rho$ $(0 \leq \rho < 1)$ and $0 \leq \lambda \leq 1$, then in $D$ we have

$$\left[\frac{f(z)}{z}\right]^{\frac{\lambda}{1-\rho}} \left[\frac{1}{2(1-\rho)}\left(\frac{zf'(z)}{f(z)} + 1 - 2\rho\right)\right]^{2(1-\lambda)} < \frac{K(z)}{z},$$

where

$$K(z) = z(1-z)^{-2}.$$

THEOREM 2.4. If $f(z) \in S^*_\rho$, then

$$f(z) = z \exp\left\{-\frac{1-\rho}{\pi}\int_0^{2\pi} \log(1-ze^{-i\theta})d\alpha(\theta)\right\}.$$

These theorems play an important role in the study of functions of $S^*_\rho$. $\alpha \geq 0$ and $0 \leq \beta < 1$ being two numbers, denote by $J(\alpha,\beta)$ the class of functions $f(z) = z + \sum_{n=2}^{\infty} a_n z^n$ holomorphic in $D$ and satisfying the conditions

$$\frac{f(z)f'(z)}{z} \neq 0, \quad \mathrm{Re}\left\{(1-\alpha)\frac{zf'(z)}{f(z)} + \alpha\left(1 + \frac{zf''(z)}{f'(z)}\right)\right\} > \beta$$

in $D$. In $J(\alpha,\beta)$ the functions

$$K_\beta(0,z) = \frac{z}{(1-z)^{2(1-\beta)}}, \quad K_\beta(\alpha,z) = \left[\frac{1}{\alpha}\int_0^z \zeta^{\frac{1}{\alpha}-1}(1-\zeta)^{-\frac{2(1-\beta)}{\alpha}} d\zeta\right]^\alpha, \quad \alpha > 0$$

play the role of extremal functions.

The class $J(\alpha,0)$ was studied by Miller and others [75, 76, 77] since 1973. In 1983, Liu [78] obtained for the functions of $J(\alpha,\beta)$, the principle of subordination:

$$\left[\frac{f(z)}{z}\right]^{\frac{1-\alpha}{2}} f'(z)^{\frac{\alpha}{2}} < \frac{1}{(1-z)^{1-\beta}},$$

the inequalities

$$-K_\beta(\alpha,-r) \le |f(z)| \le K_\beta(\alpha,r), \quad -K'_\beta(\alpha,-r) \le |f'(z)| \le K'_\beta(\alpha,r)$$

for $|z| = r < 1$, and other results.

In the same year, Ma [80] also obtained a principle of subordination:

$$\frac{zf'(z)}{f(z)} < \frac{zK'_\beta(\alpha,z)}{K_\beta(\alpha,z)}$$

and various results concerning the functions of $J(\alpha,\beta)$.

## III. SYMMETRIC UNIVALENT FUNCTIONS

$p$ being a positive integer, a function $f(z) \in S$ is said to be $p$-fold symmetric, if

$$f(\varepsilon_p z) = \varepsilon_p f(z), \quad \varepsilon_p = e^{\frac{2\pi i}{p}}.$$

The class of those functions of $S$ which are $p$-fold symmetric is denoted by $S_p$. We have $S_1 = S$ and $S_2$ is the class consisting of the odd functions of $S$. For $f(z) \in S_p$ we have

$$f(z) = z + \sum_{n=1}^{\infty} a_{np+1}^{(p)} z^{pn+1}. \tag{3.1}$$

Szegö [53] conjectured

$$|a_n^{(p)}| = 0\ (n^{\frac{2}{p}-1}), \quad p = 1,2,\ldots .$$

This conjecture is true, when $p = 1,2,3$. Yet Littlewood [55] showed that it fails to hold for sufficiently large $p$. Pommerenke [54] further proved that it is untrue for $p \ge 12$.

Chen [1] proved

$$|a_n^{(2)}| < e^2. \tag{3.2}$$

Kung [56] and then Liu [57] successively improved (3.2) to

$$|a_n^{(2)}| < 2.56 \quad \text{and} \quad |a_n^{(2)}| < 1.36. \tag{3.3}$$

Milin [58] further improved (3.3) to

$$|a_n^{(2)}| < 1.17.$$

Chen [1] also proved

$$n^{\frac{1}{3}}|a_n^{(3)}| < e^3. \tag{3.4}$$

Kung [56] and Liu [57] improved (3.4) to

$$n^{\frac{1}{3}}|a_n^{(3)}| < 6.10 \quad \text{and} \quad n^{\frac{1}{3}}|a_n^{(3)}| < 1.644$$

respectively. Levin proved

$$n^{\frac{1}{2}}|a_n^{(4)}| < A_1 \log n, \quad n^{\frac{1}{2}}|a_n^{(p)}| < A_2\sqrt{\log n} \qquad (p > 4), \tag{3.5}$$

where $A_1$ and $A_2$ are absolute constants. Yamaguti [60] found that in (3.5) one can take

$$A_1 = 5.6\cdots, \quad A_2 = 7.38\cdots . \tag{3.6}$$

Liu improved (3.6) to $A_1 = 2.31$ and $A_2 = 4.198$.

For the coefficients of the first terms in (3.1), we have

$$\left|a_{p+1}^{(p)}\right| \le \frac{2}{p}, \quad \left|a_{2p+1}^{(p)}\right| \le \frac{2}{p} e^{-2\frac{p-1}{p+1}} + \frac{1}{p} .$$

Rosenblat [62] found $|a_7^{(2)}| < 1.15$. Liu improved it to $|a_7^{(2)}| < 1.09$ and Ma [64] further improved it to $|a_7^{(2)}| < 1.053$. For functions of $S_2$ with real coefficients, Leeman proved

$$\left|a_7^{(2)}\right| < \frac{1090}{1083} < 1.0065. \tag{3.7}$$

Jiang [66] went a step further to prove

$$\left|a_{3p+1}^{(p)}\right| \le \frac{2}{3p} + \frac{8(p+1)^6}{3p^3(3p^2+3p+1)^2} . \tag{3.8}$$

In particular when $p = 2$, (3.8) reduces to (3.7).

Recently, Liu [96] proved that, for $p \ge 2$, we have

$$\left|a_{3p+1}^{(p)}\right| \le \frac{4(A^2+BC)^{\frac{3}{2}}}{9\sqrt{3}\ ACp^3} ,$$

where $A = 3p + 2$, $B = p^2 + 3p + 2$, $C = 3p^2 + 3p + 1$. In particular for $p = 2,3$, we get

$$|a_7^{(2)}| < 1.05293, \qquad |a_{10}^{(3)}| < 0.58994.$$

In his work [73] on the class $S_\rho^*$ defined above, Wu proved

THEOREM 3.1. If $f(z) \in S_p \cap S_\rho^*$, then

$$|f(z) + zf'(z)| \geq r(1+r^p)^{-\frac{2(1-\rho)}{p}} \left[1 + \frac{1-(1-2\rho)r^p}{1+r^p}\right].$$

In particular when $p = 1$, $\rho = 0$, we get

$$|f(z) + zf'(z)| \geq \frac{2r}{(1+r)^3}.$$

Let $f(z) = z + \sum_{n=1}^{\infty} a_{np+1}^{(p)} z^{pn+1} \in S_p$ and the partial sum

$$\sigma_n^{(p)}(z) = z + \sum_{\nu=1}^{n} a_{p\nu+1}^{(p)} z^{p\nu+1}.$$

Szegö proved that all the partial sums $\sigma_n^{(1)}(z)$ are univalent in the disk $|z| < \frac{1}{4}$, where the constant $\frac{1}{4}$ cannot be replaced by a larger one. Koritzky conjectured that all the partial sums $\sigma_n^{(2)}(z)$ are univalent in the disk $|z| < \frac{1}{\sqrt{3}}$, where the constant $\frac{1}{\sqrt{3}}$ cannot be replaced by a larger one. This conjecture was proved by Kung who also proved that all $\sigma_n^{(3)}$ are univalent in $|z| < \frac{\sqrt[3]{3}}{2}$. Jiang and Du [71] proved that $\sigma_n^{(p)}$ are univalent in $|z| < \sqrt[p]{\frac{p}{2(p+1)}}$ for $p = 4,5$. Hu, Pan [103, 104] and Ye [105] proved $\sigma_n^{(p)}$ are starlike in $|z| < \sqrt[p]{\frac{p}{2(p+1)}}$ for $p = 1,2,3,4,5$.

On the other hand, Kung [102] proved that $\sigma_n^{(2)}$ is univalent in $|z| < \left(1 - \frac{4 \log n}{n}\right)^{\frac{1}{2}}$ when $n > 8$. Liu Shu-qin proved that $\sigma_n^{(p)}$ is univalent in $|z| < \left(1 - \frac{4 \log n}{n}\right)^{\frac{1}{p}}$ for $p = 1,2,\ldots,1000$. Ye [105] proved that when $f(z) \in S_p \cap S^*$, $\sigma_n^{(p)}$ is starlike in $|z| < \sqrt[p]{\frac{p}{2(p+1)}}$.

## IV. SOME SPECIAL CLASSES

Frideman [67] proved that there are only nine functions of the class $S$ whose coefficients are rational integers. They are

$$z, \quad \frac{z}{1\pm z}, \quad \frac{z}{1\pm z^2}, \quad \frac{z}{1\pm z+z^2}, \quad \frac{z}{(1\pm z)^2}. \tag{4.1}$$

Shah [68] went a step further to prove that there are forty-five functions of the class $S$ whose coefficients are integers of the quadratic complex field. They are the nine functions in (4.1) and the functions:

$$\frac{z}{1\pm iz}, \quad \frac{z}{1\pm \omega z}, \quad \frac{z}{1\pm \omega^2 z}, \dots \qquad (\omega = \frac{1+\sqrt{3}\, i}{2}).$$

It is known [8] that if $k(t)$ is a function continuous in the interval $0 \le t < \infty$ with possible exception of a finite number of points of discontinuity of the first kind and is such that $|k(t)| = 1$ $(0 \le t < \infty)$, then the solution $w = f(z,t)$ of the differential equation

$$\frac{\partial w}{\partial t} = -w \frac{1+k(t)w}{1-k(t)w}$$

satisfying the initial condition $w\big|_{t=0} = z$, has the following properties:

$$f(0,t) = 0, \quad f'_z(0,t) = e^{-t},$$

for fixed $t$, $f(z,t)$ is a holomorphic and univalent function in $D$, and the limit function

$$f(z) = \lim_{t\to\infty} e^t f(z,t)$$

exists with $f(0) = 0$, $f'(0) = 1$.

Kung [69] studied some special cases of the function $k(t)$ and obtained the following results:

1) If $k(t)$ is real, then $f(z) = z/(1-c_2z+z^2)$, where $c_2$, $|c_2| \le 2$, is a real number.

2) If $k(t)$ is pure imaginary, then $f(z) = z/(1-c_2z-z^2)$, where $c_2$, $|c_2| \le 2$, is a pure imaginary number.

3) If $k(t) = e^{i\theta}p(t)$, where $\theta$ is a real constant and $p(t)$ is a real function, then $f(z) = z/(1-c_2e^{i\theta}z+e^{2i\theta}z^2)$, where $c_2$, $|c_2| \le 2$ is a real number.

Liu Shu-qin obtained similar results for the case of p-fold symmetric univalent functions.

## V. MEROMORPHIC UNIVALENT FUNCTIONS

Consider a function $f(z)$ meromorphic and univalent in $D$ and distinguish two cases:

1) $f(z)$ has a pole at $z = 0$. In this case, we may write

$$f(z) = \frac{1}{z} + \sum_{n=0}^{\infty} c_n z^n$$

which becomes a function of the class $\Sigma$ in setting $z = 1/\zeta$.

2) $f(z)$ has a pole at $z = q$ $(q \neq 0)$. In this case, without loss of generality, we may suppose that $0 < q < 1$. Furthermore, we suppose that

$$f(0) = 0, \quad f'(0) = 1.$$

The class of functions $f(z)$ satisfying all these conditions will be denoted by $S(q)$. The function

$$K_q(z) = z(1-q^{-1}z)^{-1}(1-qz)^{-1}$$

has many properties of extremal function.

Jenkins [92] proved

THEOREM 5.1. Let $f(z) \in S(q)$ and

$$f(z) = z + \sum_{n=2}^{\infty} B_n z^n, \qquad K_q(z) = z + \sum_{n=2}^{\infty} B_n(q) z^n \tag{5.1}$$

for $|z| < q$. Then we have (by de Branges theorem [141])

$$|B_n| \leq B_n(q),$$

where

$$B_n(q) = \frac{q^n - q^{n-1}}{q - q^{-1}} = \frac{1+q^2+\cdots+q^{2n-2}}{q^{n-1}}.$$

Kirwan and Schober [93] obtained the following results:

1) Let $\Phi(t)$ be a convex and nondecreasing function in $(-\infty,\infty)$. Let $f(z) \in S(q)$ and $r \in (0,1)$. Then

$$\int_{\pi}^{\pi} \Phi(+\log|f(re^{i\theta})|)d\theta \qquad \int_{\pi}^{\pi} \Phi(\pm\log|K_q(re^{i\theta})|)d\theta,$$

where the equality holds only when $r \neq q$, $f = K_q$ and $\Phi$ is one of certain particular convex functions.

In particular, selecting $\Phi(t) = e^{\lambda t}$, we have

2) $\int_{\pi}^{\pi} |f(re^{i\theta})|^{\lambda} d\theta \leq \int_{-\pi}^{\pi} |K_q(re^{i\theta})|^{\lambda} d\theta$, where $f(z) \in S(q)$, $r \in (0,1)$, $-\infty < \lambda < \infty$ and the equality holds only when $r \neq q$, $f = K_q$.

3) $\max_{|z|=r} |f(z)| \leq K_q(r)$, where $f(z) \in S(q)$, $r \in (0,1)$.

4) Let $f(z) \in S(q)$. Then $f(z)$ maps $D$ onto a domain containing the disk $|w| < q/(1+q)^2$. $q(1+q)^{-2}$ is the Koebe constant for the class $S(q)$.

5) Let $f(z) \in S(q)$. Then

$$|B_n| \leq \min\left(\frac{1+q^n}{q^{n-1}(1-q^2)}, \frac{en}{2q^{n-1}}\right),$$

where $B_n$ is defined by (5.1).

Zhang [97] obtained the following results:

THEOREM 5.2. Let $f(z)$ be holomorphic in $D$ except at $z = q$ $(0 < q < 1)$ which is a pole of $f(z)$ and suppose that $f(0) = 0$, $f'(0) = 1$. Then in order that $f(z)$ is univalent, it is necessary and sufficient that

$$\frac{1}{\pi}\iint_{|z|<1}\left|\frac{f'(z)f'(\eta)}{[f(z)-f(\eta)]^2} - \frac{1}{(z-\eta)^2}\right|^2 d\sigma_z \leq \frac{1}{(1-|\eta|^2)^2},$$

where $\eta \neq q$.

THEOREM 5.3. Let $f(z) \in S(q)$, $|z| = p < q$, then

$$\frac{(q-r)^2(1-qr)^2}{q^2(1-r^2)^3} \leq |f'(z)| \leq \frac{q^2(1-r^2)}{(q-r)^2(1-qr)^2},$$

$$|\arg f'(z)| \leq \log \frac{q^2(1-r^2)^2}{(q-r)^2(1-qr)^2}$$

Ma [142] also obtained a necessary and sufficient condition for $f(z) \in S(q)$ and other results.

## VI. TYPICALLY REAL FUNCTIONS

A function $f(z)$ holomorphic in $D$ is said to be typically real, if it satisfies the condition

$$(\operatorname{Im} z)(\operatorname{Im} f(z)) \geq 0.$$

We denote by $T_r$ the class of the typically real functions $f(z)$ such that $f(0) = 0$, $f'(0) = 1$.

Rogosinski [116] is the first to study such a class of functions. Then it is also studied by Goluzin [117] and others.

Let $f(z) = z + \sum_{n=2}^{\infty} a_n z^n \in T_r$. Robertson [118] and Goluzin [117] proved the following formula independently of one another.

$$f(z) = \frac{1}{\pi}\int_0^{\pi} \frac{z\,d\alpha(\theta)}{1-2z\cos\theta+z^2}, \tag{6.1}$$

where $\alpha(\theta)$ is a nondecreasing function in $0 \leq \theta \leq \pi$ with $\int^{\pi} d\alpha(\theta) = \pi$.

Conversely, (6.1) represents a function of $T_r$.

The class of functions $f_p(z) = z + \sum_{n=1}^{\infty} a_{pn+1}^{(p)} z^{pn+1} \in T_r$ is denoted by $T_r^{(p)}$. Hu [119] proved

$$f_2(z) = \frac{1}{\pi}\int_0^{\pi} \frac{(1+z^2)z\,d\alpha(\theta)}{1-2z^2\cos 2\theta+z^4} \in T_r^{(2)}. \tag{6.2}$$

Liu [143] proved

$$f_p(z) = \frac{1}{\pi}\int_0^{\pi} \frac{\left(1+\frac{\sin(p-1)\theta}{\sin\theta}z^p\right)z\,d\alpha(\theta)}{1-2z^p\cos p\theta+z^{2p}} \in T_r^{(p)} \qquad p = 1,2,\dots .$$

By the formula (6.1), Goluzin [117] obtained estimates of the upper bounds of $|f(z)|$, $|f'(z)|$, of $\arg f(z)$, $\arg f'(z)$, and of the coefficients of the expansion of $f(z)$. Also by the formula (6.1), Shu [120] obtained estimates of the lower bounds of $|f(z)|$, $|f'(z)|$. Corresponding estimates for functions of $T_r^{(2)}$ were obtained by Hu [119] in basing upon the formula (6.2). The class of typically real functions and some of its generalizations were also studied by Xu [121] and Chang [122, 123, 124], and various results were obtained.

## VII. RESULTS ON SUBORDINATION

Let $F(z)$ be holomorphic and univalent in $D$ and maps $D$ onto a domain $G$. A function $f(z)$ holomorphic in $D$ is said to be subordinate to $F(z)$ in $D$, if $f(0) = F(0)$ and in $D$ the values of $f(z)$ belong to $G$. Symbolically we write $f < F$. Biernaki [134] proved that if

$$f < F, \quad f \not\equiv F, \quad \arg f'(0) = \arg F'(0), \tag{7.1}$$

then $|f(z)| < |F(z)|$ for $0 < |z| < \frac{1}{4}$. Goluzin [135] improved the radius $\frac{1}{4}$ to $0.35$ and conjectured that the best value of the radius is $(3-\sqrt{5})/2 = 0.38\cdots$. This conjecture was proved by Shah [136] in showing that under the conditions (7.1) we have $|f(z)| \le |F(z)|$ for $0 < |z| < (3-\sqrt{5})/2$ and the value $(3-\sqrt{5})/2$ of the radius is the best.

On the other hand, Goluzin [135] proved that under the conditions (7.1) we have $|f'(z)| < |F'(z)|$ for $0 < |z| < 0.12\cdots$ and conjectured that, instead of the value $0.12\cdots$, the best value of the radius is $3-\sqrt{8} = 0.17\cdots$. This conjecture was also proved by Shah.

Let $f(z) < F(z)$, $f(z) = \sum_{n=0}^{\infty} b_n z^n$, $F(z) = \sum_{n=0}^{\infty} a_n z^n$. Rogosinski [138] proved the inequality

$$|b_n| \le \max(|a_1|, |a_2|, \ldots, |a_n|)$$

for $n = 1,2$. Shah [139] proved

$$|b_3| \le \sqrt{2}\max(|a_1|, |a_2|, |a_3|),$$

where the constant $\sqrt{2}$ cannot be diminished.

## VIII. UNIVALENT FUNCTIONS IN MULTIPLY CONNECTED DOMAINS

Shah [125] studied univalent functions in doubly connected domains and obtained interesting results.

Let $G$ be a doubly connected domain. If $G$ can be mapped to the domain $1 < |w| < R$, we say that $G$ belongs to $\sigma_R$. Some results of Shah are the following theorems.

THEOREM 8.1. Let $G$ be a doubly connected domain belonging to $\sigma_R$ and let the components of its complementary set be consisted of two continuous sets $C_1$ and $C_2$. Suppose that $C_1$ contains the origin and $n$ vertices

$$a_\nu = r_1 \exp\left( i\,\frac{2\pi\nu+\xi}{n}\right) \qquad (\nu=1,2,\ldots,n)$$

and that $C_2$ contains $\infty$ and $n$ vertices

$$b = r_2 \exp\left( i \frac{2\pi\mu+\eta}{n} \right) \qquad (\mu=1,2,\dots,n),$$

where $\xi,\eta$ are real numbers. Then

$$\frac{r_1}{r_2} \le \sqrt[n]{k}\,, \tag{8.1}$$

where $k$ is the positive root of the equation

$$R = \exp\left\{ \frac{\pi}{n} k \frac{P(k)}{P(\frac{1}{k})} \right\}, \quad P(k) = \int_0^1 \frac{dx}{\{x(1-x)(x+k)\}^{1/2}}.$$

The estimate (8.1) is sharp in which the equality holds, if and only if $G$ is the plane with slits consisting of the $2n$ line segments

$$z = r \exp\left( i\frac{2n\nu+\xi}{n} \right), \qquad 0 \le r \le r_1$$

$$(\nu=1,2,\dots,n).$$

$$z = r \exp\left( i\frac{(2\nu+1)\pi+\eta}{n} \right), \qquad r_2 \le r \le \infty$$

THEOREM 8.2. Let $G$ be a doubly connected domain belonging to $\sigma_R$ and let its complementary set be consisted of two continuous sets $C_1, C_2$, where $C_1$ contains the origin and $C_2$ contains $\infty$. From the origin draw $n$ rays $\ell_\nu$ $(\nu=1,2,\dots,n)$ making equal angles with each other. Denote by $d_\nu^{(1)}$ the maximum distance from the origin for $\ell_\nu \cap C_1$, and by $d_\nu^{(2)}$ the minimum distance from the origin for $\ell_\nu \cap C_2$. When the rays $\ell_\nu$ $(\nu=1,2,\dots,n)$ rotate about the origin, let $\rho(G)$ be the lower bound of $\max\limits_{1\le\nu\le n} d_\nu^{(2)} / \min\limits_{1\le\nu\le n} d_\nu^{(1)}$. Then when $G$ varies in $\sigma_R$, the lower bound $k = \rho_R^{(n)}$ of $\rho(G)$ is uniquely determined by the equation

$$R = \exp\left\{ \frac{\pi}{\sqrt[n]{k^{n-1}}} \; \frac{P\,\frac{1}{K^n-1}}{P(k^n-1)} \right\}, \qquad k > 1.$$

Loewner [126] parametric representation and Shiffer [127] Goluzin [128] method of variation are powerful tools in the study of univalent functions in simply connected domain. Kufarev [129] and Lebedev [130] extended these

these two methods to the case of doubly connected domain. Young [132, 133] gave extensions of these methods to the case of domain of connectivity n and obtained various results.

REFERENCES

1. Chen K.K., Proc. Imp. Acad. Jap., 9(1933), 465-467.

2. Chen K.K., Tohoku Math. J., 40(1935), 160-174.

3. Koebe P., Nachr. Kgl. Ges. Wiss. Göttingen, Math. phys. Kl. 1907, 191-210.

4. Koebe P., Nachr. Kgl. Ges. Wiss. Göttingen, Math. phys. Kl. 1909, 68-76.

5. Bieberbach, Math. Ann. 77(1916), 153-172.

6. Bieberbach, Sitzgsber. Preuss. Akad. Wiss. (1916), 940-955, Zweiter Halbband.

7. Gronwall, T.H., Ann. of Math. 16(1914), 72-76.

8. Goluzin, Geometric theory of functions of a complex variable. (1969) Amer. Math. Soc. p. 119.

9. Shah Tao-shing, Science Record, 4(1951), 209-212.

10. Shah Tao-shing, Science Record, 4(1951), 351-362.

11. Goluzin G., Rec. Math. (Mat. Sbornik), 21(1947), 83-117.

12. Shah Tao-shing, Science Record. 4(1951), 363-368.

13. FitzGerald, Arch. rat. Mech. Anal. 46(1972), 356-368.

14. Kung Sheng, Scientia Sinica, (1979), 237-246.

15. Kung Sheng, On FitzGerald inequalities.

16. Hu Ke, J. Jiang Xi Normal Institute, (1979), No. 1, 5-14.

17. Hu Ke, J. Jiang Xi Normal Instit., (1979), No. 2, 9-17.

18. Hu Ke, Chin. Ann. of Math. 1(1980), 421-427.

19. Hu Ke, Scientia Sinica, (1981), 141-148.

20. Ren Fu-yao, Fu-Tan University J. (Natural Science), (1979), 69-75.

21. Ren Fu-yao, Chin. Ann. of Math. 4B(4) (1983), 425-441.

22. Kung Sheng, Acta. Math. Sinica, 3(1953), 208-212.

23. Chin Yuan-shun, Acta. Math. Sinica, 4(1954), 81-86.

24. Li Ji-min, Acta. Math. Sinica, 14(1964), 367-378.

25. Li Ji-min, Some notes on the argument of the coefficients of univalent functions, (to appear).

26. Silverman, H. Silvia E.M. and Telage D.N., Pacific J. Math. 77(1978), 557-563.

27. Littlewood, Proc. Lond. Math. Soc., 23(1924), 481-519.

28. Goluzin G.M., Math. Sb., 22 : 64(1948), 373-380.

29. Milin, Math. Sb., 28 : 70(1951), 359-400.

30. Ren Fu-yao, Fu-tan University J. (Natural Science), 1(1981), 1-14.

31. Horowitz, Proc. Amer. Math. Soc. 71(1978), 217-221.

32. Kung Sheng, Scientia Sinica, 1979, 237-246.

33. Hu Ke, Proc. Amer. Math. Soc. 87(1983), 487-492.

34. Pederson, Arch. Rat. Mech. Anal. 31(1968/69), 331-352.

35. Kung Sheng, Scientia Sinica, (1979), 1157-1170.

36. Ehrig, Math. Zeits., 140(1974), 111-126.

37. Bishouty, Math. Zeits., 149(1976), 183-187.

38. Kung Sheng, Scientia Sinica, Special Issue (1979), (I), 202-213.

39. Hu Ke, Deng Sheng-nan, Ye Zhong-qiu, Kexue Tongbao, 28(1983), 189.

40. Hu Ke, Deng Sheng-nan, Ye Zhong-qiu, Jiangxi Normal Univ. J. 2(1979), 9-16.

41. Ren Fu-yao, Scientia Sinica, Special Issue (1979), (I), 275-280.

42. Garabedian, Ross and Schiffer, J. Math. Mech. 14(1965), 975-989.

43. Garabedian and Schiffer, Arch. Rat. Mech. Anal., 26(1967), 1-32.

44. Bombieri, Invent. Math. 4(1967), 27-67.

45. Liu Shu-qin, J. Northwest University (China), (1982), No. 1, 22-30.

46. Schiffer, Bull. Soc. Math. France 66(1938), 48-55.

47. Garabedian and Schiffer, Ann. of Math. (2) 61(1955), 116-136.

48. Springer, Trans. Amer. Math. Soc. 70(1951), 421-450.

49. Kubota, Kodai Math. Sem. 28(1977), 253-261.

50. Ren Fu-yao, Fu-tan Univ. J. No. 4, (1978), 93-96.

51. Ren Fu-yao, Kexue Tongbao, Vol. 25(1980), No. 4, 277-278.

52. Yao Bi-yun, Chih. Ann. of Math. (1983), 85-88.

53. Komatu, Y. Theory of conformal mapping (1944), pp. 437-507.

54. Pommerenke, Invent. Math., 3(1967), 1-15.

55. Littlewood, Quart. J. Math. (1938), 14-20.

56. Kung Sheng, Sci. Sinica, (1955), 359-373.

57. Liu Shu-qin, Advances in Math., 7(1964), 223-227.

58. Milin, Dokl. Akad. Nauk SSSR, 176(1967), 1015-1018.

59. Levin, Math. Zeits, 38(1934), 306-311.

60. Yamaguti, K.F., Mathematics (Japan) Vol. 2(1951), 82-84.

61. Liu Shu-qin, J. Northwest University (China) (1958), No. 1, 1-8.

62. Rosenblat, Rev., Ci Lima, 40(1938), 177-179.

63. Liu Shu-qin, J. Northwest University (China) (1957), No. 1, 19-33.

64. Ma Wan-cang, On the coefficients of odd univalent functions (to appear).

65. Leeman, Duke Math. J., 43(1976), 301-307.

66. Jiang Nan-ning, The estimate of the fourth coefficient of symmetric univalent functions (to appear).

67. Frideman, Duke Math. J., 13(1946), 171-177.

68. Shah Tao-shing, J. Chin. Math. Soc., 1(1951), 98-107.

69. Kung Sheng, Acta Math. Sinica, V. 3(1953), 225-230.

70. Liu Shu-qin, J. Northwest University (China), (1980), No. 4, 15-22.

71. Liu Shu-qin, Advance in Math. V.3(1957), 325-334.

72. Wu Zhuo-ren, Acta Math. Sinica, V.6 (1956), 476-489.

73. Wu Zhuo-ren, Trans. Amer. Math. Soc. 38(1964), 277-284.

74. Wu Zhuo-ren, Acta Math. Sinica, (1984), No. 2.

75. Miller, S.S., Mocanu P.T. and Reade M.O., Proc. Amer. Math. Soc. 37(1973), 553-554.

76. Miller, S.S., Proc. Amer. Math. Soc., 38(1973), 311-313.

77. Miller, S.S., Mocanu P.T. and Reade M.O., Mathematica, 20(43), (1978), No. 1, 25-30.

78. Liu Li-quan, Acta Math. Sinica, 26(1983), 179-186.

79. Juneja, O.P. and Mogra, M.L., Rev. Roum. Math. Pur. et Appl., 13(1978), No. 5, 751-765.

80. Ma Wan-cang, On 2-starlike functions of order $\beta$ (to appear).

81. Li Ji-min, The family $F^+$ of univalent functions and the effect of the convexity of the argument. Pure and app. Math. (China) (1984), V.1, No. 1.

82. Goluzin G. Rec. Math. (Math. Sbornik) 19(1946), 183-202.

83. Hayman W.K., J. Lond. Math. Soc. 38(1963), 228-243.

84. Hu Ke, New proof of Hayman's theorem, Kexue Tongbao (to appear).

85. Hu Ke, Successive coefficients of univalent functions (to appear).

86. Ilina L.P., Mat. Zamatki, 4(1968), 715-772.

87. Grinspan A.Z., Sib. Inst. Math. Novosibirsk, (1976), 41-45. (Russian).

88. Hu Ke, On the successive coefficients of univalent functions, Proc. Amer. Math. Soc. (to appear).

89. Ye Zhong-qiu, On successive coefficients of univalent functions. Jiangxi Shiyuah Xue Bao, (to appear).

90. Duren P.L., J. London Math. Soc. 19(1979), 448-450.

91. Leung Yuk, Proc. Amer. Math. Soc. 76(1979), 86-94.

92. Jenkins, Michigan Math. J., 9(1962), 25-27.

93. Kirwan and Schober, J D'anal. Math., 30(1977), 330-348.

94. Fenchel, Preuss. Akad. Wiss. Phys.-Math. Kl., 22/23(1931), 431-436.

95. Baernstein II, Acta Math. 133 : 3 - 4(1974), 139-169.

96. Liu Shu-qin, The estimate of the fourth coefficient of symmetric univalent functions (to appear).

97. Zhang Yu-lin, J. Northwest university (1980) No. 1.

98. Szegö G., Math. Ann. 100(1928), 188-211.

99. Levin V., Jahr. deutsch. Math. Vereinig, 42(1933), 68-70.

100. Koritzky G.V., Math. Sb 36(1929), 91-98.

101. Kung Sheng, Science Record, 4(1951), 333-341.

102. Kung Sheng, Acta. Math. Sinica, 4(1954), 105-111.

103. Hu Ke and Pan Yi-fei, J. Math. Resear. Expos. 4(1984), 41-44.

104. Hu Ke and Pan Yi-fei, The starlike radius of the partial sum of odd univalent functions, Chin. Ann. of Math. (to appear).

105. Ye Sen-shu, The starlike radius of the partial sum of univalent functions (to appear).

106. Ye Sen-shu, Jiangxi Normal university J., (1984), No. 1.

107. Ren Fu-yao, Chin. Ann. of Math. 4B(4) (1983), 425-441.

108. Bazilevich I.E., Math. ser. N.S., 74(1967), 133-146.

109. Hu Ke, Chin. Ann. of Math. 3 : 3(1982), 293-302.

110. Li Jiang-fan and Zhu Hui-lin, Nature Journ. 10, Vol. 6 (1983), 793-794.

111. Szegö G., Jahresber D.M.V. 31(1922), 42-43.

112. Lavrentieff and Chepeleff, Math. 2(1937), 319-326.

113. Shah Tao-shing, Fu-tan university J. 2(1956), 125-132.

114. Shah Tao-shing, Acta. Math. Sinica, No. 3 (1957), 421-432.

115. Chang Ming-yong, Advance in Math., (1955), 337-391.

116. Rogosinski W., Math. Zeits. 35(1932), 93-121.

117. Goluzin, G.M., Math. 27(1950), 201-218.

118. Robertson, M.S., Bull. Amer. Math. Soc. 41(1935), 565-572.

119. Hu Chia-kan, Acta. Math. Sinica, Vol. 6 (1956), 651-664.

120. Shu Shao-pen, Acta. Math. Sinica, V. 6 (1956), 313-319.

121. Xu Zheng-fan, Fu-tan university J. (1957), 155-171.

122. Chang Kai-ming, Fu-tan university J. (1956), 17-28.

123. Chang Kai-ming, Acta. Math. Sinica, V.8 (1958), 12-21.

124. Chang Kai-ming, Meromorphic Typically real functions in the ring.

125. Shah Tao-shing, Acta. Math. Sinica, V.6 (1956), 598-616.

126. Loewner K., Math. Ann., 89(1923), 103-121.

127. Shiffer M., Proc. London Math. Soc. 44(1938), 432-449.

128. Goluzin G.M. Math. §6, 19(1946), 203-236.

129. Kufarev and Cemuhina, Dokl. Akad. Nauk SSSR., 107(1956), 505-507.

130. Lebedev, Dokl. Akad. Nauk SSSR., 103(1955), 767-768.

131. Shlionsky, Vestnik Leningrad Univ., 13(1958), 64-83.

132. Young Wei-qi, Acta. Math. Sinica, 24(1981), 26-35.

133. Young Wei-qi, J. Beijing Institute of Technology, No. 1 (1981), 1-8.

134. Biernacki M., Mathematica, 12(1936), 49-64.

135. Goluzin G.M., Math. 29(1951), 209-224.

136. Shah Tao-shing, Science Record V.1 (1957), 201-204.

137. Shah Tao-shing, Science Record V.1 (1957), 301-304.

138. Rogosinski, W., Proc. London Math. Soc. 48(1943), 48-82.

139. Shah Tao-shing, Acta. Math. Sinica, 8(1958), 408-412.

140. Hu Ke, Jiangxi Normal Univ. J. No. 1 (1981), 1-4.

141. FitzGerald H., and Pommerenke, Ch. The de Branges theorem on univalent functions (to appear).

142. Ma Wang-cang, Schiffer differential equations and its applications of meromorphic functions (to appear).

143. Liu Shu-qin, Northwest Univ. J. No. 3 (1980), 8-10.

Department of Mathematics
Northwest University
Xian

Department of Mathematics
Peking University
Beijing

Contemporary Mathematics
Volume 48, 1985

# QUASICONFORMAL MAPPINGS

Cheng-Qi He and Zhong Li

## 1. INTRODUCTION

This article is a summary of the works on quasiconformal mappings in China.

Chinese mathematicians started to study the theory of quasiconformal mappings at the end of the 1950's. At that time, there were seminars held at Peking University and Fudan University. The works of Ahlfors [1], Lavrent'yev [26], Bers [4], Vekua [54] and many other authors were studied in these seminars. Under the influence of these works, many papers on quasiconformal mappings appeared in Chinese Journals. It was unfortunately interrupted from 1965-1976. Since 1976 the research on this branch has resumed. Chinese mathematicians interested in this field are now studying the new achievements and developments of quasiconformal mappings in the new world. Their works concern mainly the following subjects.

Parametric representation and some estimations (§2);

Compactness and existence theorems (§3);

Nonlinear elliptic systems and quasiconformal mappings (§4);

Extremal problems for quasiconformal mappings with given boundary values (§5).

We start with some notations and basic conceptions of quasiconformal mappings.

Let $D$ be a domain in the extended plane $\hat{C} = C \cup \{\infty\}$ and $Q$ a topological quadrilateral, $\overline{Q} \subset D$. Denote by $\mathrm{Mod}(Q)$ the conformal modulus of $Q$. An orientation preserving homeomorphism $f$ of $D$ is called quasiconformal, if

$$K_D[f] = \sup_{\overline{Q} \subset D} \frac{\mathrm{Mod}(f(Q))}{\mathrm{Mod}(Q)} < \infty,$$

where $K_D[f]$ is called the maximal dilatation of $f$. This geometrical definition is suggested by A. Pfluger [41]. We sometimes need to consider mappings $f$ with $K[f] \leqq K$ ($K$ = const.). Such a mapping $f$ is called a $K$-q.c. mapping.


0271-4132/85 $1.00 + $.25 per page

L. Bers [5] noted the connection between q.c. mappings and the Beltrami equation. Every q.c. mapping $w = f(z)$ is an $L^2$ generalized homeomorphic solution of a Beltrami equation

$$\partial_{\bar{z}} w - \mu(z)\partial_z w = 0,$$

where $\mu(z)$ is a measurable function $\|\mu\|_\infty < 1$. Conversely, every $L^2$ generalized homeomorphic solution of a Beltrami equation must be a q.c. mapping.

If $w = f(z)$ is a q.c. mapping, then we call $\mu_f(z) = \partial_{\bar{z}} f / \partial_z f$ the complex dilatation of f. It is easily seen that

$$K_D[f] = \operatorname*{ess\,sup}_{z \in D} \frac{1+|\mu(z)|}{1-|\mu(z)|}.$$

We also need the conception of local dilations which is defined as

$$K_f(z) = \lim_{r \to 0} \operatorname*{ess\,sup}_{|\zeta - z| < r} \frac{1+|\mu(z)|}{1-|\mu(z)|}.$$

Obviously, $K_D[f] = \operatorname*{ess\,sup}_{z \in D} K_f(z)$.

As an effort by many authors, it is known that the analytic characters of K-q.c. mappings are as follows: (i) f is absolutely continuous on lines; and (ii) f satisfies

$$|\partial_z f| + |\partial_{\bar{z}} f| \le K(|\partial_z f| - |\partial_{\bar{z}} f|)$$

for almost everywhere.

In the discussion of the relation between the analytical definition and geometrical definition, Menchoff's theorem [37] is very important which says that if a homeomorphism $w = w(z)$ of a domain D possesses the following property: for almost each point z, there are two non-collinear rays $t_1(z)$ and $t_2(z)$ starting from z such that

$$\varlimsup_{\substack{\zeta \to z \\ \zeta \in t_1 \cup t_2}} \frac{|w(z)-w(z)|}{|\zeta - z|} < \infty,$$

then $w = w(z)$ is differentiable for almost all $z \in D$.

Here we should mention the work of Chen Huai-Hui [15] in which Menchoff's theorem is generalized and he proved the following theorem: Let $w = w(z)$ be a homeomorphism of a domain D onto $\Delta$. Suppose that $D = H_1 \cup H_2 \cup D'$ where $H_1$ is at most denumerable and $H_2$ has measure zero. If for any point

$z \in H_2$, there are two noncollinear rays $t_1(z)$ and $t_2(z)$ starting from $z$ such that

$$\overline{\lim_{\substack{\zeta\to z \\ \zeta\in t_1 \cup t_2}}} \frac{|w(\zeta)-w(z)|}{|\zeta-z|} < \infty,$$

and if for any point $z \in D'$ there are three noncollinear rays, $t_1(z)$, $t_2(z)$ and $t_3(z)$, starting from $z$ such that the limit

$$\lim_{\substack{\zeta\to z \\ \zeta\in t_1 \cup t_2 \cup t_3}} \frac{|w(\zeta)-w(z)|}{|(\zeta-z)+q(z)(\overline{\zeta}-\overline{z})|}$$

exists, then $w = w(z)$ is a K-q.c. mapping. Here $q(z)$ is a given function with $|q(z)| \leq (K-1)/(K+1)$.

## 2. PARAMETRIC REPRESENTATION AND SOME ESTIMATIONS OF QUASICONFORMAL MAPPINGS

2.1. Lowner's method in the theory of univalent functions is well known. Its main idea is to express a class of functions as solutions of a differential equation with a certain initial condition. In 1959, this method was successfully generalized to the theory of q.c. mappings by Xia Dao-Xing [57], who gave a parametric representation of q.c. mappings. Later Li You-Cai [29] and Peng Cheng-Lian [42] improved the parametric representation so that it is in the best form.

THEOREM 1. Let $q(z,t)$ be a function defined in $\{(z,t)\,|\,|z| < 1,\ 0 \leq t \leq 1\}$ with the following conditions:

i) $|q(z,t)| \leq q_0 < 1$, $q_0$ = const.

ii) for any fixed $t \in [0,1]$, $q(x,t)$ is a measurable function of $z$ in $\{z\,|\,|z| < 1\}$, and for any fixed $z \in \{z\,|\,|z| < 1\}$, $q(z,t)$ satisfies the Lipschitz condition with respect to $t$:

$$|q(z,t_1) - q(z,t_2)| \leq M|t_1-t_2|$$

where $M$ is a constant. Let $w = f(z,t)$ be a q.c. mapping of the unit disc onto itself with the complex dilation $q(z,t)$ and with the conditions: $f(0,t) = 0$ and $f(1,t) = 1$ (for all $t$; $0 \leq t \leq 1$). Then $w = f(z,t)$ satisfies the differential equation:

$$\frac{\partial w}{\partial t} = \frac{w(1-w)}{\pi} \iint_{|\zeta|<1} \left[\frac{\psi(\zeta,t)}{\zeta(1-\zeta)(w-\zeta)} + \frac{\overline{\psi(\zeta,t)}}{\overline{\zeta}(1-\overline{\zeta})(1-w\overline{\zeta})}\right] d\sigma_\zeta$$

where

$$\psi(\zeta,t) = \frac{q'_t(\zeta,t)}{1-|q'_t(\zeta,t)|^2} e^{-2i\arg\partial_\zeta f^{-1}(\zeta,t)}.$$

THEOREM 2. Let $w = f(z)$ be a K-q.c. mapping of the unit disc onto itself keeping the points $z = 0$ and $z = 1$ fixed. Then there exists a function $\phi(z,t)$ which is measurable in $\{(z,t) \mid |z| < 1,\ 0 \leq t \leq \log K\}$ and satisfies the inequality $|\phi(z,t)| \leq 1/2$, such that the solution $w = f(z,t)$ of the differential equation

$$\frac{\partial w}{\partial t} = \frac{w(1-w)}{\pi} \iint_{|\zeta|<1} \left[\frac{\phi(\zeta,t)}{\zeta(1-\zeta)(w-\zeta)} + \frac{\overline{\phi(\zeta,t)}}{\overline{\zeta}(1-\overline{\zeta})(1-w\overline{\zeta})}\right] d\sigma_\zeta$$

with the initial condition $f(z,0) = z$ is equal to $f(z)$ for $t = \log K$, that is, $f(z, \log K) = f(z)$.

Analogously with the above theorems, the parametric representation of q.c. mappings of other domains (such as C, half-planes, rectangles, rings and triangles) is considered in [29], [58] and [12].

2.2. As an application of the parametric representation, Xia Dao-Xing [57] got the following important estimation:

THEOREM 3. [57]. Let $w = f(z)$ be a K-q.c. mapping of the unit disc onto itself which keeps the points $z = 0$ and $z = 1$ fixed. Then we have

$$\frac{|f(z)-z|}{\log K} \leq \frac{\Gamma^4(1/4)}{4\pi^2}.$$

This estimation is sharp.

This result is an improvement of the estimation by Lavrent'yev [26] and Belinskiy [8].

By making use of the parametric representation, one may get many other estimations. For example, Peng Cheng-Lian [42] treated K-q.c. mappings of the unit disc onto itself keeping the points $z = 1$, $\exp(2\pi i/3)$ and $\exp(4\pi i/3)$ fixed. He obtained the following estimation

$$\frac{|f(z)-z|}{\log K} \leq \frac{\Gamma^4(1/4)}{12\pi^2}.$$

Here we should also mention a result by Wang Chuan-Fang [55]. He improved a famous theorem of Mori as follows.

THEOREM 4. [55]. Let $f(z)$ be a K-q.c. mapping of the unit disc onto itself with $f(0) = 0$. Then we have the sharp estimate:

$$4^{1-K} \cdot |z|^K \le |f(z)| \le 4^{1-1/k} \cdot |z|^{1/k}.$$

Lai Wan-Cai [25] generalized the results mentioned above. For instance, he got

THEOREM 5. [25]. Let $f(z)$ be a k-q.c. mapping of the complex plane onto itself with the points $z = 0$, $1$ and $\infty$ fixed. Then for any $r$, $0 < r < \infty$, if $|f(z)| \leqq r$ for $|z| \leqq r$, we have

$$\frac{|f(z)-z|}{\log K} \le \frac{4}{\pi} rK(\frac{1}{1+r}) \cdot K(\frac{r}{1+r})$$

where

$$K(r) = \int_0^1 \frac{dt}{\sqrt{(1-t^2)(1-rt^2)}}.$$

Xia Dao-Xing and Fan Li-Li [58] got the following result:

THEOREM 6. [58]. Let $f(z)$ be a K-q.c. mapping of the disc $\{|z| < 1\}$ onto itself keeping the points $z = 0$ and $1$ fixed. Set

$$p(a,K) = \sup_{|z_1|=r_1, |z_2|=r_2} \left\{ \frac{|f(z_1)|}{|f(z_2)|} \right\}$$

where $a = r_1/r_2$, $0 < r_1 \leqq r_2 < 1$. Then $p(a,K)$ satisfies

$$\int_0^p \frac{dx}{x\eta(x)} = \log K$$

where

$$\eta(x) = \frac{1}{2\pi} \iint_{|\zeta|<\infty} \frac{(1+x)\,d\sigma_\zeta}{|\zeta(1-\zeta)(\zeta+x)|}.$$

2.3. The family of all K-q.c. mappings with a certain normalization condition is a normal family. For such a normal family, He Cheng-Qi gave some sharp estimations:

THEOREM 7. [20]. Let $f(z)$ be a K-q.c. mapping of $\{|z| < 1\}$ with the following conditions:

$$f(0) = 0, \quad |f(z)| < 1.$$

Then

$$\prod_{k=1}^{n} \frac{|f(re^{\frac{2k\pi}{n}i})|}{r^{1/k}} < 4^{1-1/k}.$$

THEOREM 8. [20]. Let $f(z)$ be a K-q.c. mapping of $\{|z| < 1\}$ satisfying the conditions:

$$f(0) = 0, \quad f(re^{\frac{2k\pi}{n}i}) = r^{1/k} e^{\frac{2k\pi}{n}i}$$

$$(k = 1,2,\ldots,n).$$

Then (1) there exist $n$ rays with the initial point $w = 0$ which divides the plane into $n$ sectors with angle $2\pi/n$ such that the length $\lambda_k$ of the segment of the $k$-th ray covered by the maximal star region $D_f^*$ in the image domain of the mapping $f$ satisfies

$$\frac{1}{n} \sum_{k=1}^{n} \lambda_k > 4^{\frac{1-K}{nK}}.$$

(2) The length $\tilde{\lambda}_k$ of the segment covered by $D_f^*$ of the ray which starts from the origin and passes the point $r^{1/k} \exp(2k\pi i/n)$ satisfies

$$\prod_{k=1}^{n} \tilde{\lambda}_k > 4^{1/K-2}.$$

The estimations in Theorems 7 and 8 are sharp. Theorem 7 is an improvement of Mori's theorem, while Theorem 8 is a generalization of the covering theorems of Goluzin and Szegö. In the paper [20], these three problems are solved in a uniform way.

It is of interest to characterize the distortion of a class of q.c. mappings by the local dilatation. He Cheng-Qi established a theorem on the distortion of conformal modulus.

THEOREM 9. [21]. Suppose $w = f(z)$ is a q.c. mapping of domain $D$, whose local maximal dilatation is denoted by $K(z)$. Then for any ring domain $R = \{z | r_1 < |z| < r_2\}$, $R \subset C$, we have

$$\int_{r_1}^{r_2} \frac{dr}{\frac{r}{2\pi}\int_0^{2\pi} K(z_0+re^{i\theta})\,d\theta} \le \operatorname{Mod} f(R) \le \frac{1}{2\pi} \int_0^{2\pi} \left( \frac{d\theta}{\int_{r_1}^{r_2} \frac{K(z_0+re^{i\theta})}{r}\,dr} \right)^{-1}.$$

This theorem has an application to the study of the degenerate Beltrami equation. (See §3 in this paper).

Lehto [28] has an interesting result: Let $f(z)$ be a q.c. mapping of a unit disc onto itself with $f(0) = 0$ and

$$\frac{1}{2\pi}\int_0^{2\pi} d\theta \int_0^1 \frac{K(re^{i\theta})-1}{r}\, dr = M < \infty.$$

Under the condition that $\partial_{\bar z} f(0) = 0$, we have the estimates:

$$e^{-M} \leq |\partial_z f(0)| \leq e^{M}$$

which are sharp.

Zhang Guang-Ho [59] gave a simple proof of this result and generalized Szegö's covering problem in the following form: for any $n$ rays starting from the origin with the same angles, there is at least one ray such that its segment covered by the image domain has the length $\geqq 4^{-1/n} e^{-M}$.

Liu Li-Quang [36] generalized Lehto's and Zhang's results for the more general case.

2.4. The variational formulas for q.c. mappings are closely related to various extremal problems. Belinskiy [9], Krushkal [23] and other authors have made extensive investigations. The first approximate expression for q.c. mappings the maximal dilatations of which approach to zero was given by Belinskiy. In his result, the maximal dilatation satisfies some special conditions. Cheng Bao-lung [11] improved his result and obtained a better form.

THEOREM 10. [11]. Suppose $q(z)$ is a bounded measurable function in the unit disc $\{|z| < 1\}$ and $|q(z)| \leqq q_0 < \varepsilon$. Let $w = f(z)$ be a q.c. mapping of $\{|z| < 1\}$ onto itself keeping the points $z = 0$ and $1$ fixed. Then

$$f(z) = z + \frac{z(1-z)}{\pi}\iint_{|\zeta|<1}\left[\frac{q(\zeta)}{\zeta(1-\zeta)(\zeta-z)} + \frac{\overline{q(\zeta)}}{\bar\zeta(1-\zeta)(1-z\bar\zeta)}\right]d\sigma_\zeta$$
$$+\, O(\varepsilon^{5/4}), \quad d\sigma_\zeta = d\xi d\eta, \quad \zeta = \xi + i\eta.$$

THEOREM 11. [11]. Suppose $q(z)$ is a measurable function and $|q(z)| \leqq q_0 < \varepsilon$. Denote by $w = f(z)$ a q.c. mapping of the complex plane with fixed points $z = 0, 1,$ and $\infty$. Then

$$f(z) = z + \frac{z(1-z)}{\pi}\iint_{|\zeta|<\infty}\frac{q(\zeta)}{\zeta(1-\zeta)(\zeta-z)}\, d\sigma_\zeta + O(\varepsilon^{5/4}),$$

$$\forall z : |z| \leq M,$$

or

$$f(z) = z + \frac{z(1-z)}{\pi} \iint_{|\zeta| \le R} \frac{q(\zeta)}{\zeta(1-\zeta)(\zeta-z)} d\sigma_\zeta + O(\varepsilon^{5/4}),$$

$$\forall z : |z| \le M, \quad R = \varepsilon^{-3/8}.$$

These formulas have applications to the extremal problem for q.c. mappings (see [24]).

Zhu Hung [60] got an integral expression of the solutions of Beltrami equations.

At the end of this paragraph, we remark that variational formulas are connected with parametric representations for q.c. mappings by variational formulas. Thus far this is the simplest proof of parametric representations.

## 3. COMPACTNESS AND EXISTENCE THEOREMS

3.1. Let $M$ be a family of homeomorphisms. We say $M$ is compact if for any sequence $\{w_n\}$ in $M$ whose definition domains have a non-degenerate kernel, there exists a subsequence $\{w_{n_k}\}$ and numbers $a_k$ and $b_k$ such that $a_k w_{n_k} + b_k$ is convergent uniformly in any compact subset of the kernel domain. For a given compact family of mappings, we add all possible limit mappings. Such a proceeding is called completing the family.

Obviously, a family of K-q.c. mapping in a domain is compact and the family obtained as a result by completing is still a family of K-q.c. mappings.

Let $w(z)$ be a homeomorphism of a domain $D$ and $D_0$ a subdomain of $D$. Denote $\Omega_0 = w(D_0)$. If $w = U(z)$ and $\zeta = V(w)$ are conformal mappings of $D_0$ and $\Omega_0$ respectively, we call the composition mapping $V \circ W \circ U^{-1}$ a subconformal equivalent mapping of $w$. It is easily seen that any subconformal equivalent mapping of a K-q.c. mapping is a K-q.c. mapping.

Let $D_1$ and $D_2$ be domains with a curve $\gamma \subset \partial D_1 \cap \partial D_2$. Suppose that $w_k$ is a homeomorphism of $D_k$ $(k=1,2)$, $w_1(D_1) \cap w_2(D_2) = \phi$. If for any $\zeta \in \gamma$ we have

$$\lim_{\substack{z\to\zeta \\ z\in D_1}} w_1(z) = \lim_{\substack{z\to\zeta \\ z\in D_2}} w_2(z),$$

then we call the function

$$w(z) = \begin{cases} w_1(z), & z \in D_1; \\ w_2(z), & z \in D_2; \\ \lim\limits_{\substack{z\to\zeta \\ z\in D_1}} w_1(z), & \zeta \in \gamma \end{cases}$$

a continuation of $w_1$ or $w_2$.

A continuation of a K-q.c. mapping is still a K-q.c. mapping.

Suppose that $w(z)$ is an orientation-preserving homeomorphism. Starting with $w(z)$, we consider all the subconformal equivalent mappings of $w(z)$ and their continuations.

Then we complete the family. Repeating this process again and again, we get a family which is denoted by $M_w$ and call it a family generated by w. He Cheng-Qi [22] proved the q.c. mappings possess the following compactness character.

An orientation-preserving homeomorphism $w(z)$ is a q.c. mapping if and only if the family $M_w$ generated by w is compact.

Let $w(z)$ be a K-q.c. mapping. Then $M_w$ is a family of K-q.c. mappings. There is an interesting question: whether $M_w$ contains all of q.c. mappings or not? He Cheng-Qi [22] gave an answer to this question:

THEOREM 12. [22] If $w(z)$ is a K-q.c. mapping of a domain then any other K-q.c. mapping is contained in $M_w$.

3.2. Consider the Beltrami equation

$$\partial_{\bar{z}} w - q(z)\partial_z w = 0$$

where $q(z)$ is a measurable function with the condition:

$$|q(z)| \leqq q_0 < 1.$$

This condition is essential for the existence of solutions. Globally it is impossible to give up the condition. However locally it can be weakened. Based on the distortion theorem [21], Tang Tung-Hao [53] got the following results:

THEOREM 13. [53]. Suppose that $q(z)$ is a measurable function in a domain D with $|q(z)| \leqq q_0 < 1$. Denote

$$p(z) = \lim_{r\to 0} \operatorname*{ess\,sup}_{|\zeta - z|\le r} |q(z)|$$

and

$$K(z) = \frac{1+p(z)}{1-p(z)}.$$

If $K(z)$ satisfies the following conditions

i) $\iint_D K(z)dxdy < \infty$;

ii) $\lim_{\delta\to 0} \int_\delta \frac{dr}{\frac{r}{2\pi}\int_0^{2\pi} K(z+re^{i\theta})d\theta} = +\infty, \quad \forall z \in D;$

iii) the set $E = \{z \in D : K(z) = \infty\}$ is a $\Sigma$ finite linear point set;

then the Beltrami equation $\partial_{\bar z} w - q(z)\partial_z w = 0$ has a generalized homeomorphic solution.

3.3. An elliptic system of equations

$$\begin{cases} -v_y + a_{11}u_x + a_{12}u_y = 0, \\ v_x + a_{21}u_x + a_{22}u_y = 0 \end{cases}$$

$$(a_{11} > 0,\ 4a_{11}a_{22} - (a_{12}+a_{21})^2 \geq \delta_0 > 0)$$

can be written in a complex form

$$\partial_{\bar z} w - q_1(z)\partial_z w - q_2(z)\overline{\partial_z w} = 0$$

$$(|q_1| + |q_2| \leq q_0 < 1)$$

where $w = u + iv$ and $z = x + iy$. This equation is a generalization of the Beltrami equation.

Obviously, every $L^2$ generalized homeomorphic solution of this equation is a q.c. mapping. We call such a solution a mapping with two complex dilatations $q_1$ and $q_2$.

It is of interest to study the problem: for a given pair $(q_1,q_2)$ where $|q_1| + |q_2| \leqq q_0 < 1$, find a q.c. mapping with complex dilatations $(q_1,q_2)$ which maps a given domain onto a canonical domain. Boyarskiy [7] considered the case of simply connected domains and proved that this problem has unique solution if the q.c. mapping in question is normalized.

It should be noted that the problem of the existence of q.c. mappings with two complex dilatations is more complicated than the case of one complex dilatation. In his proof, Boyarskiy used Schauder's fixed point theorem.

For the case of multiply connected domains, S. Parter [40] and Li Zhong [31] proved there is a q.c. mapping with a given pair of complex dilatations which maps a given domain onto the complex plane C with slits. If one requires that the canonical domains are domains bounded by circles, the problem becomes more difficult, because in this case one cannot resolve this problem. Li Zhong [30] considered a space of complex pairs $(q_1,q_2)$ and reduced the question to one of the existence of fixed point under a mapping from the space into itself.

THEOREM 14. [30] Suppose that $q_1(z,w)$ and $q_2(z,w)$ are defined in $D \times \Delta$, where $D$ is a $(n+1)$-connected domain and $\Delta = \{w : |w| < 1\}$. If $q_1$ and $q_2$ satisfy the following conditions:

i) $|q_1(z,w)| + |q_2(z,w)| \leqq q_0 < 1$, $q_0 = \text{const.}$

ii) for almost all points $z \in D$ fixed, $q_1(z,w)$ and $q_2(z,w)$ are continuous functions of $w$, then there exists an $L^2$ generalized solution $w = w(z)$ of the equation

$$\partial_{\bar z} w - q_1(z,w)\partial_z w - q_2(z,w)\overline{\partial_z w} = 0$$

which maps $D$ onto a domain bounded by $\{|z| = 1\}$ and $n$ circles in $\Delta$.

This theorem was generalized to the case in which the equation is nonlinear.

## 4. QUASICONFORMAL MAPPINGS AND NONLINEAR ELLIPTIC SYSTEMS

4.1. It is well-known that Lavrent'yev studied q.c. mappings as homeomorphic solutions of nonlinear elliptic systems:

$$\begin{cases} \phi_1(x,y,u,v,u_x,u_y,v_x,v_y) = 0 \\ \phi_2(x,y,u,v,u_x,u_y,v_x,v_y) = 0. \end{cases} \tag{**}$$

Following Lavrent'yev, we introduce the following notations:

$$E = u_x^2 + u_y^2, \quad F = u_x v_x - u_y v_y, \quad G = v_x^2 + v_y^2,$$

$$V^\beta = J_w^{-1}(z)(E \sin^2\beta - 2F \sin\beta\cos\beta + G\cos^2\beta)^{1/2},$$

$$W^\beta = J_w^{-1}(z)(V^\beta)^{-1},$$

$$\operatorname{Sin} \theta^\beta = (V^\beta)^{-1}(E\cos^2\beta + 2F\cos\beta\sin\beta + G\sin^2\beta)^{1/2},$$

$$\operatorname{tg}\alpha^\beta = (-V_x\cos\beta + u_x\sin\beta)(V_y\cos\beta - u_y\sin\beta)^{-1},$$

where $\beta \in (0,2\pi)$. In terms of $W^\beta$, $V^\beta$, $\theta^\beta$ and $\alpha^\beta$, the equation (**) can be written as follows:

$$W^\beta = F_1(x,y,u,v,V^\beta,\alpha^\beta),$$

$$\theta^\beta = F_2(x,y,u,v,V^\beta,\alpha^\beta),$$

where $F_j^\beta (j = 1,2)$ are continuously differentiable with respect to $V^\beta$ and $\alpha^\beta$. If $\theta^\beta$ and $W^\beta$ satisfy the conditions

$$0 < \theta^\beta < \pi, \qquad 0 < \frac{\partial W^\beta}{\partial V^\beta}$$

for all $\beta \in (0,2\pi)$, then we call the equation (**) a $\Lambda$-elliptic system. If there is a constant $\delta > 0$ such that

$$\delta \le \theta^\beta \le \pi - \delta, \qquad \delta \le \frac{\partial W^\beta}{\partial V^\beta}$$

for all $\beta \in (0,2\pi)$, then we call (**) a $\Lambda$-strong elliptic system.

Moreover, equation (**) is called a $\pi$-elliptic system if

$$\frac{D(\phi_1,\phi_2)}{D(u_x,v_x)} \cdot \frac{D(\phi_1,\phi_2)}{D(u_y,v_y)} - \left[\frac{D(\phi_1,\phi_2)}{D(u_x,v_y)} + \frac{D(\phi_1,\phi_2)}{D(u_y,v_x)}\right]^2 > 0$$

If there are positive constants $d$ and $K$ such that $d \le 1$ and

$$\max\left\{\left|\frac{\partial \phi_j}{\partial u_x}\right|, \left|\frac{\partial \phi_j}{\partial u_y}\right|, \left|\frac{\partial \phi_j}{\partial v_x}\right|, \left|\frac{\partial \phi_j}{\partial v_y}\right|\right\} \le K, \qquad j = 1,2,$$

then we say that equation (**) is a $\pi$-strong elliptic system. If

$$\frac{D(\phi_1,\phi_2)}{D(u_x,v_y)} \cdot \frac{D(\phi_1,\phi_2)}{D(u_y,v_x)} > 0$$

the equation is said to be of the first class.

Equation (**) can be written in the complex form:

$$\partial_{\bar z} w = g(z,w,\partial_z w)$$

where $g(z,w;\zeta)$ is differentiable with respect to $\zeta$ and there exists a constant $q_0 \leqq 1$ such that

$$|g(z,w,\zeta_1) - g(z,w,\zeta_2)| < q_0|\zeta_1 - \zeta_2|.$$

Fan Ai-Nong got the following results:

THEOREM 15. [16]. (i) With the above hypothesis, if $q_0 = 1$, the equation

$$\partial_{\bar z} w = g(z,w,\partial_z w)$$

is a $\pi$-elliptic equation of the first class, if $q_0 < 1$, it is a $\pi$-strong elliptic equation of the first class.

(ii) If $q(z,w,0) = 0$, a $\pi$-elliptic system of the first class is equivalent to a $\Lambda$-elliptic system and a $\pi$-strong elliptic system of the first is equivalent to a $\pi$-strong elliptic system.

4.2. For a simply connected domain $D$, Lavrent'yev proved that there is a solution of a $\Lambda$-strong elliptic system which maps $D$ onto the unit disc. For a given multiply connected domain, the problem was studied by Monakhov [38], Fang Ai-Nong [17] and Wu Guo-chun [56].

THEOREM 16. [17]. Let $D$ be a domain bounded by $n+1$ Jordan curves. Suppose that $g(z,w,\zeta)$ is a measurable function defined in $D \times \Delta \times C$ where $\Delta = \{|z| < 1\}$, continuous with respect to $w$ for any fixed $z \in D$ and $\zeta \in \mathbf{C}$, and satisfies the conditions:

$$g(z,w,0) = 0$$

$$|g(z,w,\zeta_1)-g(z,w,\zeta_2)| \leq q_0|\zeta_1-\zeta_2|, \quad q_0 < 1.$$

Then there exists an $L^2$ generalized homeomorphic solution of the complex equation

$$\partial_{\bar{z}} w = g(z,w,\partial_z w)$$

which maps $D$ onto a domain bounded by $\{|z| = 1\}$ and $n$ circles in $\Delta$.

Wun Gao-chun [56] considered the following equation

$$\partial_{\bar{z}} w = Q_1(z,w,\partial_z w)\partial_z w + Q_2(z,w,\partial_z w)\overline{\partial_z w}$$

where $Q_1$ and $Q_2$ satisfy the condition of elliptic type. He proved the same result for multiply connected domains.

## 5. EXTREMAL PROBLEMS FOR QUASICONFORMAL MAPPINGS WITH GIVEN CORRESPONDENCE ON THE BOUNDARIES

5.1. Let $h$ be an orientation-preserving homeomorphism of $\{|z| = 1\}$ onto $\{|w| = 1\}$ which allows a q.c. extension to $\{|z| < 1\}$. Denote by $F = F[h]$ the family of all such q.c. extensions. Then it is easily seen that there exists a mapping $w = f_0(z) \in F$ such that

$$K[f_0] = \inf_{f \in F} K[f].$$

Such a mapping is called an extremal mapping. There are some basic problems: When is the extremal mapping $f_0$ unique? When is $f_0$ a Teichmüller mapping? How large is $K[f_0]$?

K. Strebel and E. Reich systematically studied these problems. (See [45], [47], [49]). Strebel [49] established a criterion which seems to be critical in solving these problems. Let $\tilde{h}$ be an extension of $h$ to $\{1-\varepsilon \leq |z| \leq 1\}$ and $H$ the infimum of the maximal dilatations for all possible $\tilde{h}$. Strebel's criterion says that if $K_0 = K[f_0] > H$ then there is a Teichmüller mapping in $F$ associated with a quadratic differential which has finite norm. Hence, by [47], in this case, the extremal mapping is uniquely a Teichmüller mapping.

To apply Strebel's criterion, Li Zhong [32] gave the estimate of $K_0$ from below and the estimate of $H_\zeta$, the local dilatation of boundary correspondence, from above. Then he got a criterion in which the condition of the existence of extremal Teichmüller mappings is more explicit.

Let $z = e^{i\theta}$ and $w = e^{i\psi}$. Then the correspondence $h$ can be expressed in terms of $\psi$ and $\theta$:

$$\psi = \mu_h(\theta): \quad R \to R.$$

Since $h$ has q.c. extension to $|z| < 1$, $\mu_h(\theta)$ is a quasisymmetric function. For any interval $I = (a,b)$, we define

$$\rho[I] = \inf\{\rho \,|\, \rho^{-1} \leq \frac{\mu_h(x+t)-\mu_h(x)}{\mu_h(x)-\mu_h(x-t)} \leq \rho, \ \forall x - t, x, x + t \in I\}$$

and

$$\rho(\theta) = \inf_{I \ni \theta}\{\rho[I]\}.$$

On the other hand, by a fractional linear transformation, one may map the unit disc onto the upper half plane. The correspondence $h$ becomes an orientation-preserving $\mu$ of $R$ onto itself.

Set

$$\rho_0 = \inf\{\rho \,|\, \rho^{-1} \leq \frac{\mu(x+t)-\mu(x)}{\mu(x)-\mu(x-t)} \leq \rho, \ \forall x, t \in R, t \neq 0\}.$$

Let $p(\rho)$ be the conformal modulus of the quadrilateral of the upper half plane with vertexes $-1$, $0$, $\rho$ and $\infty$, the first side of which is chosen in a way such that $p(\rho)$ is an increasing function of $\rho$. Moreover, we need the function

$$q(\rho) = 1 + \frac{1}{2\pi^2} \log^2 \rho + \frac{1}{\pi} (\log \rho)\sqrt{1 + \frac{1}{4\pi} \log^2 \rho}$$

which is the maximal dilatation of an extremal mapping of the upper half plane with the boundary correspondence

$$\begin{cases} x \mapsto x, & \text{if} \quad x \geq 0; \\ x \mapsto \rho x, & \text{if} \quad x < 0. \end{cases}$$

THEOREM 17. [32]. Suppose that $\mu_h(\theta)$ is piece-wise continuously differentiable and $\mu_h'(\theta) \neq 0$. If $q(p(\theta)) < p(\rho_0)$ for all $\theta$, then there is a Teichmüller mapping in $F[h]$ associated with a quadratic differential $\phi dz^2$ of finite norm and hence it is a unique extremal mapping.

Denote by $D(z_1,z_2,z_3,z_4)$ the cross ratio of points $z_1$, $z_2$, $z_3$ and $z_4$. Let

$$A = -\sup\{D(h(z_1),h(z_2),h(z_3),h(z_4)) \mid D(z_1,z_2,z_3,z_4) = -1\}$$

where $z_1$, $z_2$, $z_3$ and $z_4$ are on the circle $|z| = 1$. Then the above theorem can be formulated in an explicit form: If $\mu = \mu_h(\theta)$ is piece-wise smooth and its right-hand derivative $\mu_+'(\theta)$ and left-hand derivative $\mu_-'(\theta)$ satisfy the condition

$$\max(\frac{1}{5}, \frac{1}{\sqrt{A}}) \leq \frac{\mu_+'(\theta)}{\mu_-'(\theta)} \leq \min(5, \sqrt{A}),$$

then $h$ has a unique extremal extension, and it is a Teichmüller mapping.

This theorem was originally stated in a slightly different form in the paper [32]. There it was assumed that the boundary correspondence $h$ is absolutely continuous with the condition

$$0 < m \leq \mu_h'(\theta) \leq M < \infty$$

for almost all $\theta$.

This theorem has been generalized to the case in which $\mu_h'$ is allowed to have isolated zeros (see [34]).

5.2. It is of interest to know how large $K_0 = K[f_0]$ is. We may just consider $h : \mathbb{R} \to \mathbb{R}$ and the domain of the upper half plane. By the Beurling-Ahlfors theorem [10], an orientation-preserving homeomorphism $h$ has a q.c. extension to the upper half plane if and only if $h$ is $\rho$-quasisymmetric on $\mathbb{R}$, i.e.,

$$\rho^{-1} \leq \frac{h(x+t)-h(x)}{h(x)-h(x-t)} \leq \rho$$

for all $x \in \mathbb{R}$ and $t \neq 0$. If $h$ is quasisymmetric on $\mathbb{R}$, then the mapping $(x,y) \to (u,v)$ defined by

$$\begin{cases} u(x,y) = \frac{1}{2}\int_0^1 [h(x+ty) + h(x-ty)]dt, \\ v(x,y) = \frac{r}{2}\int_0^1 [h(x+ty)-h(x-ty)]dt \end{cases}$$

is a $C^1$-q.c. mapping. We call it a Beurling-Ahlfors extenstion of $h$ with parameter $r$. Beurling and Ahlfors proved

$$K[f] \le \rho^2$$

where $f$ is a Beurling-Ahlfors extension of a $\rho$-quasisymmetric function.

Actually, for a given boundary mapping $h$, the result of Beurling and Ahlfors gives an estimate for $K_0$:

$$K_0 \leq \rho^2.$$

By making use of the Beurling-Ahlfors extension, Reed [43] got the estimate

$$K_0 \leqq 8\rho.$$

Li Zhong [33] improved this result to

$$K_0 \leqq 4.2\rho.$$

Lehtinen [27] recently got a new result $K_0 \leqq 2\rho$. Without being aware of this result, Chen Ji-Xiu [14] and Tan De-Lin [52] independently obtained $K_0 < 2.58\rho$ and $K_0 \leqq 2\rho$ respectively.

Since the Beurling-Ahlfors extension is not necessarily an extremal extension, the estimates of maximal dilatations of Beurling-Ahlfors extensions may not be good estimates for $K_0$. Li Zhong [33] points out that one cannot go further to estimate the maximal dilatations by making use of Beurling-Ahlfors extensions. He proved that the maximal dilatations of Beurling-Ahlfors extensions of $\rho$-quasisymmetric functions are not smaller than $1.59\rho$ (when $\rho$ is sufficiently large) no matter how the parameter $r$ is chosen. Based on the work ([33]), Li Wei and Liu Yong [35] proved that for Beurling-Ahlfors extensions, the estimate $K \leqq 2\rho$ is sharp in the sense that the coefficient 2 cannot be smaller.

5.3. Let $\Omega$ be a domain on the plane $C$ and $w = f(z)$ a q.c. mapping of $\Omega$ with the complex dilatation $\kappa(z)$. Denote by $Q_f(\Omega)$ the set of all q.c. mappings of $\Omega$ which have the same boundary value as $f$. We consider the linear-functional operator

$$\Lambda_\kappa[\phi] = \iint_\Omega \kappa(z)\phi(z)dxdy$$

where $\phi$ is regular and belongs to $L^1(\Omega)$. The family of all such functions $\phi$ is denoted by $B(\Omega)$. As usual, the norm of the operator $\Lambda_\kappa$ is defined by

$$\|\Lambda_\kappa\| = \sup_{\phi\in B(\Omega)} \left|\iint_\Omega \kappa(z)\phi(z)dxdy\right| / \|\phi\|.$$

Hamilton [19] proved that if $f$ is an extremal mapping in $Q_f(\Omega)$, then $\|\Lambda_\kappa\| = \|\kappa\|_\infty$. Reich and Strebel [47] proved that this condition is also sufficient for simply connected domains. Later Strebel [51] generalized this result to general open Riemann surfaces. So $f$ is an extremal mapping if and only if $\|\Lambda_\kappa\| = \|\kappa\|_\infty$.

Now the questions are as follows: How many extremal mappings are there in $Q_f(\Omega)$? When is the extremal mapping unique? Reich [46] propounded the question whether the uniqueness of extremal mappings is connected with the uniqueness of the Bahn-Banach extension of $\Lambda_\kappa$ from $B(\Omega)$ to $L^1(\Omega)$.

If the operator $\Lambda_\kappa$ has a unique Bahn-Banach extension from $B(\Omega)$ to $L^1(\Omega)$ and $\|\Lambda_\kappa\| = \|\kappa\|_\infty$, then we shall say that $\kappa$ belongs to HBU. We do not know whether the following statement is true: $f$ is a unique extremal Teichmüller mapping if and only if $\kappa \in$ HBU. Even we do not know whether $|\kappa(z)|$ is almost everywhere equal to a constant when $\kappa \in$ HBU.

These questions were studied by Ren Fu-Yao [48]. He obtained some positive answers to these questions with an additional condition.

THEOREM 18. [48]. (1) If $\kappa \in$ BHU and the function $|\kappa(z)|$ is continuous on a set $\Omega\setminus E$ where $E \subset \Omega$ has measure zero, then $|\kappa| = \|\kappa\|_\infty$ for almost all points in $\Omega$.

(2) If $\|\Lambda_\kappa\| = \|\kappa\|_\infty$, then the q.c. mapping $f$ with complex dilatation x is the unique extremal Teichmüller mapping and $\kappa = k\overline{\phi_0}/|\phi_0|$ ($\phi_0 \in B(\Omega)$) if and only if $|\kappa(z)|$ is continuous on a set $\Omega\setminus E$ where $E \subset \Omega$ has measure zero, $\arg \overline{\kappa(z)} = \arg \phi_0$ ($\phi_0 \in B(\Omega)$) and $\Lambda_\kappa$ has a unique Bahn-Banach extension from $B(\Omega)$ to $L^1(\Omega)$.

5.4. We are also interested in the extremal problem of q.c. mappings with given boundary values and bounds of dilatations.

Let $T$ be a measurable subset of $D = \{|w| < 1\}$ and $\overline{T} \subset D$. Let $b(w)$ be a measurable function on $T$ and satisfy

$$0 < b_0 \le b(w) \le b_1 < 1.$$

If $z = F(w)$ is a q.c. mapping of $D$ onto itself, we denote by $A(F,T,b(w))$ the family of all q.c. mappings $G$ with the conditions: (i) $G|_{\partial D} = F|_{\partial D}$; (ii) $|\partial_{\bar{w}}G| \le b(w)|\partial_w G|$, for $w \in T$. Set

$$k_G = \text{ess sup}\{|\partial_{\bar{w}}G|/|\partial_w G|\{w \in D\setminus T\}.$$

Suppose $F \in A[F,T,b(w)]$ and $f = F^{-1}$ has complex dilatation $k(z)$. If $k_F > 0$, we define

$$\tau(z) = \begin{cases} k(z)/b(f(z)), & \text{for } z \in F(T); \\ k(z)/k_F, & \text{for } z \in D\setminus F(T). \end{cases}$$

Reich has proved that $k_F = \inf_{G \in A} k_G$ if and only if $k_F = 0$ or $k_F > 0$ and $\|\Lambda_\tau\| = 1$.

Chen Ji-Xiu [13] generalized the result of Reich to a case in which $b(w)$ can assume the value 0 and 1. He assumed that $b(w)$ is a function defined on $T$ satisfying the following conditions:

i) $b(w)$ is continuous almost everywhere on $T$,

ii) $T_1 = \{w \in T | \|b(w)\|_u^\infty = 1$ for any neighborhood $U$ of $w\}$ consists of finitely many points,

iii) for any point $w_0 \in T_1$ there is a constant $M > 0$ such that

$$\frac{1}{2\pi}\int_0^{2\pi} B(w_0+re^{i\theta})d\theta < M \log\frac{1}{r}$$

for sufficiently small $r > 0$ where

$$B(w) = [1+b(w)]/[1-b(w)].$$

Suppose $z = F(w) \in A[F,T,b(w)]$ and the complex dilatation of $F^{-1}$ is $k(z)$. If $k_F > 0$, then we define

$$\tau(z) = \begin{cases} 0, & z \in F(T_0); \\ k(z)/b(f(z)), & z \in F(T\setminus T_0); \\ k(z)/k_F, & z \in D\setminus F(T), \end{cases}$$

where $T_0 = \{w \in T\setminus T_1 | b(w) = 0\}$.

THEOREM 19. [14]. With the above conditions, $k_F$ is equal to $\inf\{k_G\}$ if and only if $k_F = 0$ or $k_F > 0$ and

$$\sup_{\|\phi\|=1,\phi\in B(D\setminus T_0)} \left|\iint_{D\setminus T_0'} \tau(z)\phi(z)dxdy\right| = 1, \quad T_0' = F(T_0).$$

## REFERENCES


1. Ahlfors, L.: On quasiconformal mappings. J. Analyse Math. 3 (1953), 1-58 and 207-208.

2. Ahlfors, L.: Quasiconformal reflections. Acta. Math. 109 (1963), 291-301.

3. Ahlfors, L. and Beurling, A.: Conformal invariants and function-theoretic null-sets. Acta Math. 83 (1950), 101-129.

4. Bers, L.: On a theorem of Mori and the definition of quasiconformal mappings. Trans. Amer. Math. Soc. 84 (1957), 78-84.

5. Bers, L.: The equivalence of two definitions of quasi-conformal mappings, Comment. Math. Helv. 37, No. 2 (1962), 148-154.

6. Bers, L.: Quasiconformal mappings and Teichmüller theorems, Analysis Functions, Princeton (1960), 89-120.

7. Bojarski, B. V.: Generalized solutions of first-order system of elliptic type with discontinuous coefficients, Math. Sb. 43 (1957), 451-503.

8. Belinskiy, P. P.: Distortion under quasiconformal mappings, Dokl. Akad. Nauk SSSR, 91 No. 5 (1953), 997-998.

9. Belinskiy, P. P.: On the solution of extremal problems of quasiconformal mappings by the method of variations, Dokl. Akad. Nauk SSSR, 91, No. 2 (1958), 199-201.

10. Beurling, A. and Ahlfors, L.: The boundary correspondence under quasi-conformal mappings. Acta Math. 96 (1956), 125-142.

11. Cheng Bao-Lung: An approximate expression for E-quasiconformal mappings and extremal problems, Acta Math. Sinica (China) 14 (1964), 212-217.

12. Chen Ji-Xiu: Quasiconformal mappings of canonical domains, Fudan Xuebao, 21 (1982), 77-86.

13. Chen Ji-Xiu: Quasiconformal homeomorphism with given boundary values and bound of complex dilations, to appear.

14. Chen Ji-Xiu: On the Beurling-Ahlfors extensions, to appear.

15. Chen Huai-Hui: Absolutely continuous functions and a sufficient condition for K-quasiconformality, Shuzue Jinzhan, 7 (1964), 84-93.

16. Fang Ai-Nong: Quasiconformal mappings and function theory of nonlinear elliptic systems of first-order, Acta Math. Sinica (China) 23 (1980), 280-292.

17. Fang Ai-Nong: Riemann mapping theorem for nonlinear quasiconformal mappings, 23 (1980), 241-353.

18. Fehlmann, R.: Ueber extremal quasiconforme Abbildungen, Comm. Math. Helv. 56 (1981) 558-580.

19. Hamilton, R. S.: Extremal quasiconformal mappings with prescribed boundary values, Trans. Amer. Soc. 138 (1969), 399-406.

20. He Cheng-Qi: An estimate of the distortion of quasiconformal mappings, Scientia Sinica (China) 11 (1983), 967-974.

21. He Cheng-Qi: A theorem on the distortion of modulus under quasiconformal mappings, Acta Math. Sinica (China) 15 (1965), 487-494.

22. He Cheng-Qi: Compactness for quasiconformal mappings, Acta Math. Sinica (China) 13 (1963), 447-453.

23. Krushkal, S. L.: Variation of a quasiconformal mapping of closed Riemann surfaces, Daklady Akad. Nauk SSSR 157 No. 4 (1964), 781-783.

24. Krushkal, S. L.: Quasiconformal mappings and Riemann surfaces, W. H. Winston & Sons (1979)

25. Lai Wan-Cai: Hyperbolic metric on $C(0,1,\infty)$ and quasiconformal mappings, Proceedings of the conference of functions theory at Shanghai (1964).

26. Lavrent'yev, M. A., Sur une classe de représentations continues, Math. sb., 42, No. 4 (1935), 407-424.

27. Lehtinen, M.: The dilatation of Beurling-Ahlfors extension of quasi-symmetric functions, Ann. Acad. Sci. Fenn. Ser. A.I.Math. 8 (1983), 187-191.

28. Lehto, O., Virtanen, K. J. and Väisälä, J.: Contribution to the distortion theory of quasiconformal mappings, Ann. Acad. Sci. Fenn, Ser. AI., No. 273 (1959), 3-14.

29. Li Yu-Cai: Parametric representations of quasiconformal mappings, Shuxue Jinzhan, 9 (1966), 55-66.

30. Li Zhong: A theorem on the existence of homeomorphic solutions of a quasilinear elliptic system of first-order, Acta Math. Sinica 13 (1963), 454-461.

31. Li Zhong: Modified Dirichlet problem and its applications to quasiconformal mappings, Beida Xuebao, 4 (1964), 319-340.

32. Li Zhong: On the existence of extremal Teichmüller mappings, Comm. Math. Helv., 57 (1982), 511-517.

33. Li Zhong: On Beurling-Ahlfors extensions, Acta Math. Sinica, 26 (1983), 279-290.

34. Li Zhong: On the existence of extremal Teichmüller Mappings (II), Beijing daxue Xuebao, 6 (1982), 1-9.

35. Li Wei and Liu Yung: On maximal dilatations of Beurling-Ahlfors extensions of quasisymmetric functions, Kexue Tungsen, Vol. 29 (1984), 1151.

36. Liu Li-Qian: A remark on quasiconformal mappings of multiply connected domains, Heilungjuang Daxue Xuebao 1 (1982), 36-44.

37. Menchoff, D.: Sur les differentelles totales des functions univalentes, Math. Ann. 105 (1931), 75-85.

38. Monakhov, V. N.: Kakl. Nauk SSSR 220 (1975), 520-523.

39. Mori, A.: On quasiconformality and pseudo-analyticity, Trans. Amer. Math. Soc. 84 (1957), 56-77.

40. Parter, S.: On mappings of multiply connected domains by solutions of partial differential equations, Comm. Pure and Appl. Math. 13 (1960), 167-182.

41. Pfluger, A.: Quasiconforme Abbildungen und logarithmische Kapazität, Ann. Inst. Fourier Grenoble 2 (1951), 69-80.

42. Peng Cheng-Lian: On the parametric representation of quasiconformal mappings and extremal problems, Fudan Xuebao, 10-11 (1965-66), 15-28.

43. Read, T. J.: Quasiconformal mappings with given boundary values, Duke Math., 33 (1966), 459-464.

44. Reich, E.: On the relation between local and global properties of boundary values for extremal quasiconformal mappings, Annals of Math. Studies 79 (1974), 391-407.

45. Reich, E.: Quasiconformal mappings with prescribed boundary values and a dilatation bound. Arch. Rational Mech. Anal. 68 (1978), 99-112.

46. Reich, E.: A criterion for unique extremality of Teichmüller mappings, Indiana Univ. Math. J. 30 (1981), 441-447.

47. Reich, E. and Strebel, K.: Extremal quasiconformal mappings with given boundary values, Contributions to analysis, Academic Press, New York-London, 1974, 375-391.

48. Ren Fu-Yao: On the three conjectures of Reich, to appear.

49. Strebel, K.: On the existence of extremal Teichmüller mappings, Journal d'analyse Math. 30 (1976), 464-484.

50. Strebel, K.: On quadratic differentials and extremal quasiconformal mappings, International Congress, Vancouver, 1974.

51. Strebel, K.: On quasiconformal mappings of open Riemann surfaces, Comm. Math. Helv. 53 (1978), 301-321.

52. Tan De-Lin, Estimate of local dilatations of Beurling-Ahlfors extensions, to appear.

53. Tang Tung-Hao: Univalence of solutions of degenerate Beltrami equations, Fudan Xuebao, 10-11 (1965-66), 93-100.

54. Vekua, I. N.: Generalized Analytic Functions (International series in Pure and Applied Math. Vol. 25), New York, Pergamon, 1962 (trans. from Russian).

55. Wang Chuan-Fang: A sharp form of Mori's theorem on Q-mappings, Kexue Jilu 4 (1960), 334-337.

56. Wun Guo-Chun: A basic theorem of quasiconformal mappings, Hebei Xuagung Xueyuan Xuebao, 2 (1980), 20-40.

57. Xia Dao-Xing: A parametric representation of quasiconformal mappings, Kexue Jilu, 3 (1959), 323-329.

58. Xia Dao-Xing and Fan Li-Li: Parametric representation of quasiconformal mappings and its applications, Selected works of science and technology (Shanghai), 1959, 66-99.

59. Zhang Guang-Ho: A theorem on quasiconformal mappings, Shuxue Zhenzhan Vol. 8 (1965), 387-394.

60. Zhu Hung: An expression of homeomorphic solutions of Beltrami equations, Fudan Xuebao, 9 (1964), 23-29.

Department of Mathematics
Fudan University
Shanghai

Department of Mathematics
Peking University
Beijing

Contemporary Mathematics
Volume 48, 1985

# RIEMANN SURFACES

Ming-Yong Zhang

While an increasing interest in Riemann surfaces apparentently arose in Chinese mathematical community in the last few years, for several decades only very few Chinese mathematicians have been concerned with this topic. In my knowledge, the earliest atricle [19] on Riemann surfaces written by a Chinese appeared in 1941, in which the author Yang Ou Tchen described a method of constructing the "fewest" sheeted complete covering surface with given branch points of given order over a given Riemann surface. Since then four articles [20, 21, 22, 23] by Yang Tsung-Pan B. were published in 1943 and 1954. Yang was to comment in these papers on various classical ideas on Riemann surfaces rather than to obtain new results. In 1955, Zhang Ming-Yong in an article [26] on interior mappings extended the Morse' topological methods for plane domains [8] to Riemann surfaces and deduced some inequalties for the valency of a meromorphic function from the following general result: For a pseudo-harmonic function $u$ on a compact bordered 2-dimensional manifold $A$ whose boundary is composed of $n$ Jordan curves

$$d - a - \sum_{i=1}^{n} I_i = 2(1 - g) - v,$$

where $d$ denotes the number of logarithmic poles of $u$, $a$ the sum of multiplicites of saddle points, $I_1, \ldots, I_n$ the indice of the boundary contours respectively and $g$ the genus of $A$.

Among the diverse appraoches to Riemann surfaces the famous Ahlfor's extension of Schwarz Lemma [1] is a curious one that deserves to be called differential-geometric. In 1956 Zhang Ming-Yong [27, 28] improved this result to the following form: Let $R$ be a Riemann surface produced by a conformal mapping $w = w(z)$ from the unit disk and $\lambda(w)|dw|$ be a continuous metric with generalized Gaussian curvature $K^* = -\lambda^{-2}\Delta^*\log\lambda \leq -4$ everywhere on $R$, where $\Delta^*$ is essentially the Blaschke's generalized Laplacian made conformally invariant, then

$$\lambda(w)|dw| \leq (1 - |z|^2)^{-1}|dz|.$$


0271-4132/85 $1.00 + $.25 per page

Moreover, the equality never occurs at a point $w_0$ where $K^* < -4$. This improvement enabled him not only to determine the exact value of a constant $T_\rho$ of Bloch type, but also to prove the uniqueness of the corresponding extremal case.

Let me recall in this connection the definition of $T_\rho$. $T_\rho$ is the largest constant such that any convex domain of order $\rho$ produced by a conformal mapping $w = w(z)$ with $|w'(0)| = 1$ from the unit disk can cover a circular disk of radius $T_\rho$, where by a convex domain of order $\rho$ is meant a domain $D$ such that through each of its boundary points a circle of radius $\rho$ can be described to contain $D$ in its interior. Zhang Ming-Yong [24, 25] had proved previously by means of the Ahlfor's extension of Schwarz Lemma that $T_\rho$ is the positive root of the equation

$$\sqrt{T_\rho(2\rho - T_\rho)}\ \sin^{-1}\sqrt{\frac{T_\rho}{2\rho}} = \frac{\pi}{4} \qquad (1 \leq \rho < \infty)$$

and

$$T_\infty = \frac{\pi}{4}$$

and that the extremal case is attained when the domain is bounded by two circular arcs of radius $\rho$ (degenerating to a parallel strip when $\rho = \infty$) with $w(0)$ as its center of symmetry. The varification of the uniqueness makes, however, the fore-mentioned improvement inevitable.

After 1956 had seen in China few works on Riemann surfaces until a renewing stage initiated around 1981. Thus in 1982 and 1984 four articles by Chen Huaihui [3,4,5] and Chen Huaihui and Zhang Sunyan [6] appeared in which were given detailed discussions of Abel's theorem based on the modern foundations laid down by Ahlfor's and Sario [2]. Wu Jiongqi, following M. Nakai [10,11], investigated in his thesis [17] the elliptic dimension of a bordered Riemann surface of genus zero with null ideal boundary, where by the elliptic dimension is meant the dimension of the cone of the positive solutions, assumed to vanish continuously on the border, of the equation $\Delta u = pu$ with $p \geq 0$. By means of a compactification of Martin type Wu proved that there exists at least one minimal element over each ideal boundary component and consequently the elliptic dimension is no less than card $\delta$, where $\delta$ denotes the set of all ideal boundary components. This generalizes Nakai's result in the sense that the Nakai's additional assumption of the discreteness of $\delta$ is omitted. Wu [18] has also proved recently that the same is valid for the case of any genus. Qui Shuxi dealt in his thesis [13, 14] and other articles [15, 16] with classification theory as well as the theory of meromorphic functions on a Riemann surface of some null class. He verified that a meromorphic function on an end of a Riemann surface of

$O^o_{AD}$ is of bounded valency if and only if its cluster set at the ideal boundary is AD-removable and deduced as a consequence that a Riemann surface of $O^o_{AD}$ is essentially maximal. Later, the last assertion was generalized by Zhang Ming-Yong [29] into the following result: A Riemann surface $R$ is essentially maximal if and only if one of the following conditions is satisfied: 1) $R$ is a subsurface with locally SD-removable complementary set on a maximal surface. 2) The maximal surface into which $R$ can be imbedded conformally is unique up to a conformal homeomorphism. This result reveals itself also a generalization of Oikawa's theorem [12] which concerns the Riemann surfaces of finite genus of $O_{AD}$.

Now it seems impossible to make an exhaustive survey of current trends in the studies of Riemann surfaces in China, since quite a number of interesting articles are still to appear or in preparation. Therefore I can but content myself with sketching in concluding some works in which a number of mathematicians of Xiamen University are engaging, apart from those of Wu, Qiu and M.Y. Zhang summarized in the preceding paragraph. Zhang Xun is to give in his thesis a thorough description of the behaviou of a subharmonic function at the ideal boundary of a parabolic Riemann surface by means of potentials with a kernel which is a mixture of the modified Green functions and the kernel of Martin type. Gong Xian-Zong investigates the extensions and the singularities of a subharmonic function of a Riemann surface as well as in a harmonic space. Zhang Ming-Yong is to characterize various function-theoretic null sets on a Riemann surface in terms of complex potentials. Lin Yong in his thesis modifies based on a new topology of a Riemann surface the concept of Fuglede's finely holomorphic function so as to preserve the validity of the fundamental theorem of Calculus. At last, though may be slightly wandering from our topic, it is nevertheless of interest to mention Gao Qi-Ren's negative solution [7] of Doob-Hwang's Problem about fine limit. In 1983, J.S. Hwang [8] proved the Doob's conjecture that there exists a meromorphic function in the unit disk such that it possesses fine limit at every point on the unit circle, but every point on the unit circle is its Julia point. Hwang then asked whether "meromorphic" can be replaced by "holomorphic". Far stronger than necessary, Gao gave the general assertion that the fine lower limit of a superharmonic function in the unit disk at a point on the unit circle is never greater than the Stolz lower limit. A simple truth but overlooked by experts.

From above it is reasonable to say that a flourishing period is foreseen. We should rather report any more about it at a next time.

BIBLIOGRAPHY

[1] Ahlfors, L.V., An Extension of Schwarz Lemma, Trans. Amer. Soc. 43 (1938) 359-364.

[2] Ahlfors, L.V. and Sario, L., Riemann Surfaces, 1960.

[3] Chen Huaihui, Abel's Theorem for Compact Bordered Riemann Surfaces (in Chinese), Adv. Math. (Chinese) 11 (1982), 315-317.

[4] Chen Huaihui, A General Abel' Theorem (in Chinese), Acta Math. Sinica 25(1982), 603-609.

[5] Chen Huaihui, Able's Theorem for Open Riemann Surfaces (in Chinese), Journ. Beijing Univ.

[6] Chen Huaihui and Zhang Sunyan, Able's Theorem for Compact Bordered Riemann Surfaces, Ibid.

[7] Gao Qi-Ren, The Negative Answer to Doob-Hwang's Problem about Fine Limit and Julia Point (to appear).

[8] Hwang, J.S., On Doob's Conjecture about Fine Limit and Julia Point, Adv. Math. 48(1983), 75-81.

[9] Morse, M., Topological Methods in the Theory of Functions of a Complex Variable, 1951.

[10] Nakai, M., A Test for Picard Principle, Nagoya Math. J. 56(1974), 105-119.

[11] Nakai, M., Picard Principle for Finite Densities, Nagoya Math. J. 70 (1978), 7-24.

[12] Oikawa, K., On the Uniqueness of the Prolongation of an Open Riemann Surface of Finite Genus, Proc. Amer. Math. Soc. 11(1960), 785-787.

[13] Qiu Shuxi, Supplements to the Classification Theory of Riemann Surfaces (Chinese- English Abstract), J. Xiamen Univ. (Nat. Sc.) 20(1981) 416-425.

[14] Qiu Shuxi, On Function-theoretic Null Sets on a Riemann Surface (Chinese-English Abstract), Ibid. 21(1982), 138-145.

[15] Qiu Shuxi, Classes and of Riemann Surfaces (Chinese-English Abstract), Ibid. 22(1983), 427-432.

[16] Qiu Shuxi, The Essential Maximality and Some Other Properties of a Riemann Surface of (to appear in Chinese Ann. Math.).

[17] Wu Jiongqi, The Elliptic Martin Boundary over a Compact Set of Capacity Zero (in Chinese), Chinese Ann. Math. 4(A)(1983), 689-698.

[18] Wu Jiongqi, The Elliptic Dimension of a Bordered Riemann Surface with Null Ideal Boundary (to appear in J. Xiamen Univ.).

[19] Yang Ou Tchen, Surfaces de Riemann régulières de remification donnés, C.R. 213(1941), 556-558.

[20] Yang Tsung-Pan B., Analys zur Definition der Riemannschen Flächen, Tohoku M.J. 49(1943), 208-212.

[21] Yang Tsung-Pan B., Ein elementares Potential auf einer geschlossenen kouformen Riemannschen Fläche und seine Anwendungen (Chinese-German summary), Acta Math. Sinica 4(1954), 279-294.

[22] Yang Tsung-Pan B., Über einen Existenzsatz (Chinese-German summary), Ibid. 4(1954), 295-299.

[23] Yang Tsung-Pan B., Einige Eigenschaften nichtfortsetzbarer konformer Riemannscher Flächen (Chinese, German summary), Ibid. 4(1954), 301-304.

[24] Zhang Ming-Yong, Ein Überdeckungssatz für konvexe Gebiete. Science Record 5 (1952) 17-21.

[25] Zhang Ming-Yong, The Bloch Constant for Convex Conformal Mappings (Chinese), Shuxue Jinzhan (Adv. Math. (Chinese)) 1(1955), 387-391.

[26] Zhang Ming-Yong, Interior Mappings and Pseudoharmonic Functions (Chinese), J. Xiamen Univ. (Nat. Sc.) 1955(5), 84-92.

[27] Zhang Ming-Yong, Eine Verallgemeinerung des Laplaceschen Operators und subharmonische Funktionen (Chinese-German summary), Ibid. 1956(2), 36-43.

[28] Zhang Ming-Yong, Die Blockschen Funktionen in knovexen, Abbildungen (Chinese-German summary), Ibid. 1956 (2), 44-48.

[29] Zhang Ming-Yong, Essentially Maximal Riemann Surfaces, (Read at the 4th Assembly of Delegates of The Chinese Math. Soc., 1983).

DEPARTMENT OF MATHEMATICS
XIAMEN UNIVERSITY
Xiamen

Contemporary Mathematics
Volume 48, 1985

# APPROXIMATION AND INTERPOLATION IN THE COMPLEX DOMAIN

Xie-Chang Shen

Chinese mathematicians are involved in the research on the approximation and interpolation in the complex plan and have obtained various kinds of results. The present article is a survey of some of these results and consists of six sections namely; §1. The estimation of order of approximation by polynomials; §2. The estimation of order of approximation by rational functions; §3. The expansion and estimation of remainder by means of rational functions; §4. The incompleteness, basis and moment problem of some systems of functions; §5. The closure of $\{z^{\tau_n}\ell n^j z\}$ on unbounded curves and domains; §6. The completeness of the system $\{f(\lambda_n z)\}$.

## §1. THE ESTIMATION OF ORDER OF APPROXIMATION BY POLYNOMIALS

Let $G$ be a domain bounded by a closed rectifiable Jordan curve $\Gamma$ in the complex plane and $G_\infty$ be the complement of the closed domain $\overline{G}$. We denote by $w = \Phi(z)$, $\Phi(\infty) = \infty$, $\Phi'(\infty) > 0$, the function mapping $G_\infty$ conformally onto $|w| > 1$ and by $z = \Psi(w)$ its inverse function.

Denote by $\theta(s)$ the angle between the positive real axis and the tangent at the point $z = z(s)$ of the curve $\Gamma$ with the arc length $s$. Writing

$$j(h) = \max_{|s-s'|\le h} |\theta(s)-\theta(s')|, \tag{1.1}$$

the curve $\Gamma$ is said to be fairly smooth if

$$\int_0 \frac{j(h)}{h}|\ell n\, h|\, dh < +\infty \tag{1.2}$$

(see [1]). Alper [1] has obtained some theorems on the order of approximation by polynomials in a domain bounded by fairly smooth curve. Chen considered the approximation by Cesaro combination of Faber's polynomials.


0271-4132/85 $1.00 + $.25 per page

THEOREM 1 (Chen [2]). Let $G$ be a domain bounded by a fairly smooth curve $\Gamma$ and $f(z)$ be a function which is the uniform derivative of a bounded function, then the Cesaro mean

$$\sigma_n^\gamma(z;f) = \sum_{k=0}^{n} \frac{(\gamma)_{n-k}}{(\gamma)_n} a_k \Phi_k(z)$$

of Faber's series $\Sigma a_k \Phi_k(z)$ of $f(z)$, satisfies the inequality

$$\max_{z \in \bar{G}} |\sigma_n^\gamma(z;f) - f(z)| \le \frac{C_\gamma}{1-\alpha} \omega(\frac{1}{n+1}; f), \tag{1.3}$$

where $(\gamma)_n = (\gamma+1)(\gamma+2)\cdots(\gamma+n)/n!$, $\gamma > 0$.

$$a_k = \frac{1}{2\pi i} \int_\Gamma \frac{f[\Psi(\tau)]}{\tau^{k+1}} d\tau, \qquad k = 0,1,2,\ldots .$$

$\Phi_k(z)$ is the Faber's polynomial of degree $\le k$ (see [3]),

$$\omega(t;f) = \max_{\substack{z',z \; \bar{D} \\ |z'-z| \le t}} |f(z') - f(z)|$$

$$\alpha = \overline{\lim_{t \to 0}} \, t \frac{\omega'(;f)}{\omega(t;f)} < 1 \tag{1.4}$$

and $C_\gamma$ is a constant depending on $\gamma$ only.

Here and after we denote by $C$ or $C_1, C_2, \ldots$ constants which can take different values.

THEOREM 2 (Chen [2]). Under the conditions of Theorem 1, if $\alpha = 1$, then

$$\max_{z \in \bar{G}} |\sigma_n^\gamma(z;f) - f(z)| \le C_\gamma \omega(\frac{1}{n+1}; f) \ln(n+1). \tag{1.5}$$

THEOREM 3 (Chen [2]). Under the conditions of Theorem 1 except (1.4), if

$$\beta = \underline{\lim}_{t \to 0} \, t \frac{\omega'(t;f)}{\omega(t;f)} > 0$$

then

$$\max_{z \in \bar{G}} |\sigma_n^\gamma(z;f) - f(z)| \le C_\gamma \omega(\frac{1}{n+1}; f). \tag{1.6}$$

COROLLARY. If $f(z) \in \text{Lip}\,\alpha$ in the domain $G$ bounded by a fairly smooth curve, then for $\gamma > 0$

(1.7) $$\max_{z \in G} |\sigma_n^\gamma(z;f) - f(z)| \le \frac{C_\gamma}{n^\alpha}.$$

It is the Alper's result for $\gamma = 1$ (see [1]).

Su [4] considered the approximation for functions of class $E_1(G)$ in complex plane.

THEOREM (Su [4]). Let $G$ be a domain bounded by a fairly smooth curve $\Gamma$ and $f(z)$ be a function which has r-th derivative $f^{(r)}(z) \in E_1(G)$, then for every natural number $n$ there exists a polynomial $P_n(z)$ of degree $\le n$, satisfying the inequality

(1.8) $$\|f(z) - P_n(z)\|_{L(\Gamma)} \le \frac{C}{n^r}\,\omega\left(\frac{1}{n}; f_0^{(r)}\right),$$

where $f_0^{(r)}(w) = f^{(r)}[\Psi(w)]$.

Shen and Lou [5] investigated the problem of approximation by polynomials in a $j_\lambda$ domain.

DEFINITION. A domain is said to be a $j_\lambda$ domain, if the norm $j(h)$ of the angle between the tangent to the boundary of this domain and the positive real axis satisfies the condition

(1.9) $$\int_0 \frac{j(h)}{h} |\ln h|^\lambda dh < +\infty, \qquad 0 < \lambda \le 1$$

or the conditions

(1.10) $$\int_0 \frac{j(h)}{h} |\ln|\ln h||dh < +\infty$$

for $\lambda = 0$.

THEOREM (Shen and Lou [5]). Let $G$ be a $j_\lambda$ domain and $f(z)$ be a function which has p-th continuous derivative $f^{(p)}(z)$, $\omega(t;f^{(p)}) \le \omega_0(t)$, where $\omega_0(t)$ is a modulus of continuity of some continuous function and $\omega_0(t)|\ln t|^{1-\lambda}$ monotonously tends to zero when $t \to 0$, then for every natural number $n$, there exists a polynomial $P_n(z)$ of degree $\le n$ such that

(1.11) $$\max_{z \in G} |f(z) - P_n(z)| \le \frac{C}{n^p}\,\omega_0\left(\frac{1}{n}\right)(\ln n)^{(p+1)(1-\lambda)}.$$

There are some results on approximation in terms of means (see [6]).

Let function $f(z)$ be analytic in a simply connected bounded domain $G$ and satisfy

$$\|f(z)\|_G := \left\{\iint_G |f(z)|^p dxdy\right\}^{\frac{1}{p}} < +\infty, \qquad p \geq 1. \tag{1.12}$$

Then we write $f(z) \in H'_p(G)$ and denote by $\omega_p(t;f)$ its integral modulus of continuity:

$$\omega_p(t;f) = \sup_{|h|\leq t} \|f(z+h) - f(z)\|_G \tag{1.13}$$

where we define $f(z) = 0$, when $z \overline{\in} \overline{G}$. For the case of the unit circle $|z| < 1$ we can define the k-th integral modulus of continuity:

$$\omega_p^{(k)}(t;f) = \sup_{|h|\leq t} \|\Delta_h^k(z;f)\|_{|z|<1-k|h|}$$

where the k-th difference is defined by

$$\Delta_h^k(z;f) := \sum_{j=0}^{k} (-1)^{k-j} C_k^j f(z+jh).$$

We denote by $\rho_n^{(p)}(f;G)$ the best approximation of $f(z)$ by polynomials of degree $\leq n$ in terms of means in the domain $G$:

$$\rho_n^{(p)}(f;G) := \inf_{P_n} \|f(z)-P_n(z)\|_G \tag{1.14}$$

where the "inf" is taken over all the polynomials of degree $\leq n$.

THEOREM 1 (Xing [6]). Suppose that function $f(z)$ has r-th derivative $f^{(r)}(z) \in H'_p(|z|<1)$ and $\omega_p(t;f) \leq \omega_0(t)$, where $\omega_0(t)$ is the modulus of continuity of some continuous function. Then for every natural number $n$

$$\rho_n^{(p)}(f;|z|<1) \leq \frac{C}{n^r} \int_0^{1/n} \frac{\omega_0(t)}{t}\, dt. \tag{1.15}$$

Obviously when $\omega_0(t) \leq ct^\alpha$, $0 < \alpha \leq 1$, we get the Alper's result [7], but in the paper [7] the proof is not quite satisfactory.

THEOREM 2 (Xing [6]). Suppose that the continuous function $\Omega(u)$ is monotonously increasing on $u \geq 0$, $\Omega(0) = 0$ and that there exists a constant $c$ such that $\Omega(2u) \leq c\Omega(u)$

$$\int_0^t \frac{\Omega(u)}{u} du < +\infty.$$

Moreover, suppose that for every $n \geq 1$ there exists a polynomial $P_{2^n}(z)$ of degree $\leq 2^n$ such that

$$\|f(z) - P_{z^n}(z)\|_{|z|<1} \le \frac{C}{2^{nm}} \Omega(\frac{1}{2^n}),$$

where $m \ge 0$ is an integer. Then $f^{(m)}(z) \in H_p'(|z|<1)$ and the k-th integral modulus of continuity $\omega_p^{(k)}(t;f^{(m)})$ satisfies

$$\omega_p^{(k)}(t;f^{(m)}) \le \begin{cases} Ct^k \int_t^1 \frac{\Omega(u)}{u^{k+1}} du, & \text{when } m = 0 \\ C\left[t^k \int_t^1 \frac{\Omega(u)}{u^{k+1}} du + \int_0^t \frac{\Omega(u)}{u} du\right], & \text{when } m = 1,2,\dots . \end{cases}$$

COROLLARY. If for every $n \ge 1$ there exists a polynomial $P_{2^n}(z)$ such that

$$\|f(z) - P_{2^n}(z)\|_{|z|<1} \le \frac{C}{2^{n(m+\alpha)}}, \qquad 0 < \alpha \le 1, \quad m = 0,1,\dots,$$

then $f^{(m)}(z) \in H_p'(|z|<1)$ and

$$\omega_p(t;f^{(m)}) \le Ct^\alpha, \qquad 0 < \alpha < 1,$$

$$\omega_p^{(2)}(t;f^{(m)}) \le Ct, \qquad \alpha = 1.$$

§2. THE ESTIMATION OF ORDER OF APPROXIMATION BY RATIONAL FUNCTIONS

The problem of order of approximation by rational functions is very interesting and was investigated in several papers. Two cases are to be distinguished according to that the poles of the rational functions realizing approximation are preassigned or arbitrary [8], [9], [10].

Let $\{\alpha_i\}, |\alpha_i| > 1$, $i = 1,2,\dots$, and $\{\beta_j\}, |\beta_j| > 1$, $j = 1,2,\dots$ be two sequences of complex numbers and

$$\varepsilon_n(\alpha) = \left[\sum_{i=1}^{n} (1 - \frac{1}{|\alpha_i|})\right]^{-1}, \quad \varepsilon_n(\beta) = \left[\sum_{j=1}^{m} (1 - |\beta_j|)\right]^{-1} \tag{2.1}$$

THEOREM 1 (Shen and Lou [11]). Suppose that the function $f(z)$ is continuous on $z = 1$ and has p-th continuous derivative on $|z| = 1$. Then for every pair of natural numbers $n$ and $m$ there exists a rational function $R_{n+m}(z)$ with the prescribed poles at $\{\alpha_i\}$, $i = 1,2,\dots,n$ and $\{\beta_j\}$,

$j = 1,2,\ldots,m$:

$$(2.2) \qquad R_{n+m}(z) = \sum_{h=0}^{n+m} b_h z^h \Big/ \prod_{i=1}^{n} (z-\alpha_i) \prod_{j=1}^{m} (z-\beta_j)$$

such that

$$(2.3) \qquad \max_{|z|=1} |f(z) - R_{n+m}(z)| \le C \left\{ [\varepsilon_n(\alpha)]^p \omega_p(\varepsilon_n(\alpha); f^{(p)}) + [\varepsilon_n(\beta)]^p \omega_p(\varepsilon_m(\beta); f^{(p)}) + q^{\frac{1}{\varepsilon_n(\alpha)}} + q^{\frac{1}{\varepsilon_m(\beta)}} \right\},$$

where $\omega_p(t;f^{(p)})$ is the modulus of continuity of $f^{(p)}(z)$ on $|z| = 1$ $\varepsilon_n(\alpha)$ and $\varepsilon_m(\beta)$ are defined by (2.1), $q$ is an absolute constant, $0 < q < 1$.

The classical Jackson's Theorem is obtained in taking $\alpha_1 = \alpha_2 = \cdots = \alpha_n = \infty$, $\beta_1 = \beta_2 = \cdots = \beta_m = 0$ and $n = m$.

It is obvious that if the function $f(z)$ to be approximated is analytic in $|z| < 1$ and continuous on $|z| \le 1$, then the similar result can be obtained by using only $\{\alpha_i\}$ as the poles of the approximating rational functions.

The above theorem 1 essentially improves a result in Mergelian and Dzarbasjan's paper [12], in which the function to be approximated satisfies $f^{(p)}(z) \in \mathrm{Lip}\,\alpha$, besides there is an extra factor $\{\ell n[\varepsilon_n(\alpha) + \varepsilon_m(\beta)]\}^{p+\alpha}$ in their estimation which is therefore not precise.

I would like to indicate that our paper [11] was finished in 1964, but it was published in 1977 because of the Cultural Revolution. In 1977 similar results were obtained by Andersson and Ganelius [13] and also by Pekarskii in 1977 [14].

Shen and Lou [11] got results also in the inverse sense.

Let $\{\alpha_n\}$, $|\alpha_n| > 1$, $n = 1,2,\ldots$ be a sequence of complex numbers satisfying

$$(2.4) \qquad \sum_{n=1}^{\infty} \left(1 - \frac{1}{|\alpha_n|}\right) = +\infty$$

$$(2.5) \qquad \overline{\lim_{n\to+\infty}} \frac{\ell n \frac{1}{\Delta_n}}{\ell n \sum_{k=1}^{n} \left(1 - \frac{1}{|\alpha_k|}\right)} = \mu, \quad \Delta_n = \min_{1\le k\le n} \left(1 - \frac{1}{|\alpha_k|}\right)$$

and when $\mu \neq +\infty$

$$(2.6) \qquad \overline{\lim_{n\to+\infty}} \frac{[\varepsilon_n(\alpha)]^\mu}{\Delta_n} = \nu, \qquad \varepsilon_n(\alpha) = \left[\sum_{k=1}^{n}\left(1-\frac{1}{|\alpha_k|}\right)\right]^{-1}.$$

The sequence $\{\alpha_n\}$ is said to belong to class $W[\mu,\nu]$, if either the upper limit in (2.5) is less than $\mu$, or it is equal to $\mu$ but the upper limit in (2.6) is less than or equal to $\nu$.

THEOREM 2 (Shen and Lou [11]). Suppose that $\{\alpha_i\}$, $|\alpha_i| > 1$, $i = 1,2,\ldots$ belongs to $W[\mu,\nu]$ and $\{\beta_j\}$, $|\beta_j| < 1$, $j = 1,2,\ldots$ belongs to $W[\mu',\nu']$, $\mu < +\infty$, $\mu' < +\infty$. Let $R_{n+m}(z)$ be a rational function with its poles at $\alpha_1, \alpha_2,\ldots,\alpha_n$ and $\beta_1, \beta_2,\ldots,\beta_m$ (see (2.2)). Then for every natural number $k$

$$(2.7) \qquad \max_{|z|\le 1}|R^{(k)}_{n+m}(z)| \le C_1\left\{\left[\sum_{i=1}^{n}\left(1-\frac{1}{|\alpha_i|}\right)\right]^{2\mu+1+\varepsilon(\nu)} + \left[\sum_{j=1}^{m}(1-|\beta_j|)\right]^{2\mu'+(+\varepsilon(\nu))}\right\} \max_{|z|\le 1}|R_{n+m}(z)|,$$

where

$$\varepsilon(x) = \begin{cases} 0 \ , & x < +\infty \\ \eta \ , & x = +\infty, \end{cases}$$

$\eta > 0$ is a given small number and $C_1$ depends upon $k$, $\nu$ and $\nu'$.

The Classical Bernstain inequality can be obtained in taking $\alpha_i = \infty$, $i = 1,2,\ldots,n$, $\beta_j = 0$, $j = 1,2,\ldots,m$, $\mu = \mu' = 0$, and $\nu = \nu' = 1$.

In a certain sense, the estimate (2.7) is precise. For example if we disregard $\{\beta_j\}$ and take $\alpha_i = 1 + \frac{1}{i^\alpha}$, $0 < \alpha < 1$, $\mu = \frac{\alpha}{1-\alpha}$, then the equality in (2.7) can be attained at least for the order of magnitude.

J. Szabados was also interested in similar inequality [15].

Now we suppose that the function $f(z)$ is defined on $|z| = 1$. Let

$$(2.8) \qquad E[f;\alpha_1,\ldots,\alpha_n;\beta_1,\ldots,\beta_m] = \inf_{\{b_i\}} \max_{|z|=1}|f(z) - R_{n+m}(z)|,$$

where the "inf" is taken over all the rational functions $R_{n+m}(z)$ (2.2) with the poles of $\alpha_i$, $|\alpha_i| > 1$, $i = 1,2,\ldots,n$ and $\beta_j$, $|\beta_j| < 1$, $j = 1,2,\ldots,m$.

THEOREM 3 (Shen and Lou [11]). Suppose that $\{\alpha_i\}$ belongs to $W[\mu,\nu]$, $\mu < +\infty$ and $\{\frac{1}{\beta_j}\}$ belongs to $W[\mu',\nu']$, $\mu' < +\infty$. If for any natural numbers $n$ and $m$

$$(2.9)\quad E[f;\alpha_1,\ldots,\alpha_n;\beta_1,\ldots,\beta_m] \le C\left[[\varepsilon_n(\alpha)]^{(2\mu+1)(p+\alpha)} + [\varepsilon_m(\beta)]^{(2\mu'+1)(p+\alpha)}\right],$$

where $p \ge 0$ is an integer, $0 < \alpha \le 1$, $\varepsilon_n(\alpha)$ and $\varepsilon_m(\beta)$ are defined by (2.1), then

1. If $\nu < +\infty$, $\nu' < +\infty$, then $f(z)$ has p-th continuous derivative $f^{(p)}(z)$; moreover $f^{(p)}(z) \in \mathrm{Lip}\,\alpha$ when $0 < \alpha < 1$, and the modulus of continuity of $f^{(p)}(z)$ $\omega_p(t;f^{(p)}) = O(t|\ell n\, t|)$, when $\alpha = 1$;

2. If $\nu < +\infty$ or $\nu' < +\infty$, then $f(z)$ has p-th continuous derivative $f^{(p)}(z)$ and $\omega_p(t;f^{(p)}) \in \mathrm{Lip}(\alpha-\varepsilon)$, where $\varepsilon > 0$ is an arbitrary small number.

Theorem 3 improved the inverse theorem in Mergelian and Dzarbasjan's paper [12] in which it is assumed that no limiting point of $\{\alpha_i\}$ or $\{\beta_j\}$ is on $|z| = 1$.

Furthermore, the order of approximation by rational functions

$$Q_{n+m}(z) = \sum_{i=1}^{n} \frac{a_i}{(z-\alpha_i)^2} + \sum_{j=1}^{m} \frac{b_j}{(z-\beta_j)^2}$$

is investigated in paper [11].

Shen and Lou have got similar theorems for functions in $H^p (p \ge 1)$ spaces [16].

If the approximation is considered in more general domains the following three problems naturally arise: 1. What kind of curve should be the boundary of the domains? 2. What conditions should be posed on the poles of the approximating rational functions? 3. Can we obtain similar order of approximation as in the approximation theory on real axis? These three problems are related with each other.

Elliott [17] and Walsh [18] considered the case where the boundary of the domain is a closed analytic curve and the poles of the approximating rational functions are requested to have no concentrated points on the boundary of the domains. Obviously, these conditions are very strong. Shen and Lou [6] investigated the order of approximation by rational functions in $j_\lambda$ domains, under rather strong conditions on the distribution of the poles. They obtained the following theorem.

THEOREM (Shen and Lou [5]). Suppose that the domain $G$ is a $j_\lambda$ domain (see the definition (1.9)) and the function $f(z)$ is analytic in $G$ continu- ous on $|z| \le 1$ and has p-th continuous derivative on $|z| \le 1$, $\omega(t;f^{(p)}) \le \omega_0(t)$, where $\omega_0(t)$ is a modulus of continuity of some continuous function and $\omega_0(t)|\ln t|^{1-\lambda}$ montonously tends zero when $t \to 0$. If $\{\alpha_i\}$ belongs to $G_\infty$- the complement of the closed domain $\overline{G}$, and satisfies the condition:

$$\overline{\lim_{n\to+\infty}} \frac{\ln n}{\sum_{i=1}^{n}\left(1-\frac{1}{|\alpha_i'|}\right)} < \frac{1}{2}, \qquad \alpha_i' = \Phi(\alpha_i) \tag{2.10}$$

where $\Phi(z)$ is a mapping function (see §1), then for any natural number $n$ there exists a rational function with its poles at $\alpha_i$, $i = 1,2,\dots,n$

$$Q_n(z) = \sum_{k=1}^{n} b_k z^k \Big/ \prod_{i=1}^{n}(z-\alpha_i) \tag{2.11}$$

such that

$$\begin{aligned}&\max_{z\in G}|f(z) - Q_n(z)| \\ &\le C\left\{[\varepsilon_n(\alpha)]^p\omega(\varepsilon_n(\alpha');f^{(p)})|\ln|\varepsilon_n(\alpha')||^{(p+1)(1-\lambda)} + q^{\frac{1}{\varepsilon_n(\alpha')}}\right\},\end{aligned} \tag{2.12}$$

where $q$ is an absolute constant, $0 < q < 1$,

$$\varepsilon_n(\alpha') = \left[\sum_{i=1}^{n}\left(1-\frac{1}{|\alpha_i'|}\right)\right]^{-1}, \tag{2.13}$$

and when $p = 0$, $\lambda = 1$, the result (2.12) is still valid without any restric- tions on the distribution of the poles of the approximating rational functions.

This result improves not only the above mentioned Elliott and Walsh's result but also that of Coqialian. In his paper [19] $j_1$ domain is considered and no restriction is posed on the distribution of the poles of the approxima- ting rational functions, but there is an extra factor $\ln n$ produced by using Mergelian and Dzarbasjan's estimate [12].

The inverse theorem is also obtained in the paper [5].

In 1978 Shen [20] (see [21]) obtained the same estimate as (2.12) in $J_\lambda$ domain, but under the weaker condition than (2.10) on the distribution of the poles

(2.14) $$(1-\lambda)\overline{\lim_{n\to+\infty}}\frac{\ln\ln n}{\sum_{k=1}^{n}\left(1-\frac{1}{|\alpha'_k|}\right)} < 1, \qquad 0 \le \lambda \le 1$$

(when $\lambda = 1$, there is no restriction on the distribution of the poles).

All these approximation theorems are valid in $G_\infty$ by means of rational functions with poles in $G$ and also valid in doubly connected domain. They are valid also on closed $j_\lambda$ curve.

It is worthy to indicate that in the appendix written by Mergelian in Walsh's book [22] the approximation by rational functions with its poles outside a continuum $E$ was considered on $E$ and some condition were posed on the distribution of these poles, which is similar to our condition (2.10), but our condition is weaker than that of Mergelian. Besides he didn't investigate the order of approximation in the appendix. Under the condition (2.10) and by using the method of proof in [5] and some well-known theorems (see the first chapter of Mergelian's paper [23]) it is easy to prove the above mentioned Mergelian's result, but he himself didn't give any proof.

On the other hand, Fichera [24], [25], [26] considered the case that on boundary of the domain $f$ has continuous derivative satisfying the Lipchitz condition and the poles of the approximating rational functions have no concentrated points on the boundary. He obtained the density theorem only. Obviously this is a very special case of our results.

There are some papers on the best approximation by rational functions with fixed poles for functions in the Smirnov spaces $E_p$, $p \ge 1$.

Let the boundary $\Gamma$ of a domain $G$ be a closed smooth curve and the norm $j(h)$ of $\alpha(s)$ - angle between the tangent to $\Gamma$ and the positive real axis satisfy

(2.15) $$\int_0 \frac{j(h)}{h}\,dh < +\infty.$$

Alper obtained the order of approximation by polynomials in the domain satisfying the condition (2.15) for functions in space $E_p$, $p > 1$ [27]. Here we introduce some results on best approximation by rational functions.

THEOREM (Shen and Lou [28]). Let the boundary $\Gamma$ of a domain $G$ satisfy the condition (2.15). Suppose that the function $f(z)$ is analytic in $G$ and $f(z) \in E_p$, $p > 1$, $\{\alpha_i\}$ is a sequence of complex numbers in $G_\infty$. Then for any natural number $n$ there exists a rational function $R_n(z)$ of degree $n$ with its poles at $\alpha_i$, $i = 1,2,\dots,n$ (see (2.11)) such that

$$\|f(z) - R_n(z)\|_{L_p(\Gamma)} \le C\left\{\omega_p(\varepsilon_n(\alpha');f) + q^{\frac{1}{\varepsilon_n(\alpha')}}\right\}, \tag{2.16}$$

where $q$ is an absolute constant $0 < q < 1$ and $\varepsilon_n(\alpha')$ is defined by (2.14).

If the function $f(z)$ to be approximated has derivative of higher order belonging to $E_p$ and if we want to improve the order of approximation, then it is necessary to pose condition (2.10) on the distribution of the poles of the approximating rational functions in the paper [27]. In that paper the inverse theorem has also been studied.

In 1980, without using the condition (2.10), Shen [29] obtained better order of approximation by rational functions, in the approximation of analytic functions which together with its higher order derivative belong to $E_p$, $1 < p < +\infty$, in domains satisfying the condition (2.15).

Recently we studied a new class of domains, the $K_q$ domains, $q > 1$. These are domains $G$ such that its boundary $\Gamma$ is a closed rectifiable Jordan curve and for any $f(\xi) \in L_q(\Gamma)$, the function

$$S(z) = \frac{1}{2\pi i}\int_\Gamma \frac{f(\xi)}{\xi - z}d\xi, \qquad z \in G_\infty$$

belongs to $E_q(G_\infty)$ (see [29], [30]).

Some results have been established for the equivalence of $K_q$ domains. This kind of domain is more general then the domains mentioned above, its boundary can consist of a finite number of curves satisfying the condition (2.15) (see [31]).

THEOREM (Shen [30], [32]). Suppose that $G \in K_q$, $q > 1$, then for $f(z) \in E_p(G)$, $\frac{1}{p}+\frac{1}{q} = 1$ and any natural number $n$, there exists a rational function $R_n(z)$ with its poles at $\alpha_i$, $i = 1,2,\ldots,n$ (see (2.11)) such that

$$\|f(z) - R_n(z)\|_{L_p(\Gamma)} \le C\left\{\omega_p(\varepsilon_n(\alpha');\tilde{f}) + q^{\frac{1}{\varepsilon_n(\alpha)}}\right\}, \tag{2.17}$$

where

$$\tilde{f}(w) = \frac{1}{2\pi i}\int_{|\tau|=1} \frac{f[\Psi(\tau)]\Psi'(\tau)^{\frac{1}{p}}}{\tau - w}\,d\tau,$$

and $q$ is an absolute constant, $0 < q < 1$ and $\varepsilon_n(\alpha')$ is defined by (2.14).

REMARK: The smoothness of $\tilde{f}(w)$ on $|w| = 1$ is better than that of $f[\Psi(\tau)]\Psi'(\tau)^{\frac{1}{p}}$ on $|w| = 1$.

Su [4] obtained some results for the space $E_1(G)$, in which the domains are assumed to be $j_1$ domains and no condition is posed on the distribution of the poles of the approximating rational functions.

For the space $H_p'(|z|<1)$ Ying [6] has got the following theorem.

THEOREM (Ying [6]). Suppose that $f(z) \in H_p'(|z|<1)$ and $\omega_p(t;f) \le \omega_0(t)$, where $\omega_0(t)$ is the modulus of continuity of some continuous function. Then for any natural number $n$, there exists a rational function $R_n(z)$ with its poles at $\alpha_i$, $i = 1,2,\ldots,n$ (see (2.11)) such that

$$\|f(z) - R_n(z)\|_{|z|<1} \le C\left\{\int_0^{\varepsilon_n(\alpha)} \frac{\omega_0(t)}{t}dt + e^{-\frac{1}{10}\frac{1}{\varepsilon_n(\alpha)}}\right\}, \tag{2.18}$$

where $\varepsilon_n(\alpha)$ is defined by (2.1).

Recently, Chui and Shen [33] applied the theory of approximation by rational functions to digital filter realization and obtained some results.

Let $f(z) \in H_2(|z|<1)$, $f(z) \ne 0$, $|z| \le 1$ and the polynomial $Q_N(z)$ of degree $\le N$

$$Q_N(z) = b_0 + b_1 z + \cdots + b_N z^N \tag{2.19}$$

satisfy

$$\|1 - f(z)Q_N(z)\|_{L_2(|z|=1)} = \inf_{S_N(z)} \|1 - f(z)S_N(z)\|_{L_2(|z|=1)}, \tag{2.20}$$

where the "inf" is taken over all the polynomials $S_N(z)$ of degree $\le N$. The polynomial $Q_N(z)$ is called the N-th degree polynomial least-square inverses of $f(z)$.

In 1982 Chui and Chan [34] proved:

1. $Q_N(z) \ne 0$, $|z| \le 1$;

2. As $N \to +\infty$, the function $\frac{1}{Q_N(z)}$ tends to $f(z)$ in $H^2$ space. Hence we can consider $\frac{1}{Q_N(z)}$ as the approximate of $f(z)$ and it gives a stable all-pole filter. Moreover, Chui and Shen [33] gave the estimate of

$$\max_{|z|\le 1}\left|f(z) - \frac{1}{Q_N(z)}\right|$$ in the following theorem.

THEOREM 1 (Chui and Shen [33]). Let $f(z) \in H^2(|z|<1)$ and $\frac{1}{f(z)} \in C(a,b,c)$:

$$C(a,b,c) = \left\{ f(z) \middle| f(z) \text{ is analytic in } |z| < 1, \ f(z) \in C^\infty \text{ on } |z| \le 1 \right.$$

$$\left. \text{and } |f(e^{it})^{(r)}| \le C_1 r^a b^r r^{cr} \quad r = 0,1,2,\ldots \right\}$$

where $a \ge 0$, $b > 0$, $c > 0$ and $C_1$ is a constant, and let $Q_n(z)$ be the polynomial of degree $n$ which is least-square inverse of $f(z)$. Then

$$(2.21) \qquad \max_{|z|\le 1} \left| f(z) - \frac{1}{Q_n(z)} \right| \le K n^{\frac{a}{c}} \ell^{-\frac{c}{e\, b^{1/c}} n^{\frac{1}{c}}},$$

where K-constant indepening of $n$. Further, the exponent $\frac{1}{c}$ of $n$ cannot be improved.

REMARK. In our filter realization we take $f(z) \in C(a,b_1,4)$, where $b_1 > \frac{8}{e^4}$ and $a$ is some constant depending on $b_1$, $f(z)$ has the real Taylor coefficients and $f(e^{it})$-even function. If

$$F(z) = \frac{a_0}{2} + \sum_{k=1}^{\infty} a_k z^k, \qquad |z| < 1,$$

where $a_k = e^{-k^{\frac{1}{c}}}$, $k \ge 3$ and $a_0$, $a_1$, $a_2$ are chosen such that

1. $a_k$ monotonously tends to zero;
2. $\Delta^2 a_k = a_{k+2} - 2a_{k+1} + a_k$, $k = 0,1,2,\ldots$, then we have $F(z) \in C(1,c^c,c)$

and $f(z) = \frac{1}{F(z)} \in C^\infty$ on $|z| \le 1$. Moreover, for any polynomial $S_n(z)$ of degree $\le n$ we have

$$\max_{|z|=1} \left| f(z) - S_n(z) \right| \ge K n^{\frac{c+1}{2c}} e^{-(n+1)^{\frac{1}{c}}},$$

where $K > 0$ is a constant.

THEOREM 2 (Chui and Shen [33]). Let $H_m^R(a,b,c)$ be the collection of the functions $f(z)$ analytic in $|z| < 1 + R$ for some $R > 0$ with the exception of the line segments $\rho e^{it_j}$, $1 \le \rho \le 1 + R$, $j = 1,2,\ldots,m$ and infinitely continuously differentiable on $|z| \le 1$, with

$$|f^{(r)}(z)| \le M\, r^a b^r r^{cr}, \qquad r = 0,1,\ldots, \qquad |z| = 1$$

where $a \geq 0$, $b \geq 0$, $c > 1$ and $M > 0$ is a constant. Then for any integer $k \geq 2$ there exists a rational function $R_n(z)$ of degree $\leq n$ with its poles in $|z| > 1$ such that

$$\max_{|z|\leq 1} |f(z) - R_n(z)| \leq K e^{-\frac{n^{\frac{1}{2}}}{(\ell n\ n)^{\frac{3}{2}+\varepsilon}}}, \tag{2.22}$$

where $\varepsilon > 0$ is an arbitrary small number and $K$ is a constant depending on $a$, $b$, $c$, $\varepsilon$ only.

Szabados [35, 36] and Rusak [10], [37] both investigated the same kind of problems, but in their papers the conditions posed on the function $f(z)$ to be approximated are different from ours, the results are then different too.

In Theorem 2, the poles of the rational approximants constructed lie on many rays, and since the functions to be approximated only have $m$ "bad" points $z_j = e^{it_j}$, $j = 1,2,\ldots,m$ on the unit circle, it would be interesting to require all the poles to lie on the rays $\rho e^{it_j}$, $\rho > 1$, $j = 1,2,\ldots,m$. It should be emphasized that in digital filter realization these poles, which give rise to the feed-back coefficients, are very important. In fact, polynomial approximants which only give none-recursive digital filter are not quite useful. We have the following.

THEOREM 3 (Chui and Shen [33]). Let $f(z) \in H_m^R(a,b,c)$, where $a \geq 0$, $b > 0$ and $c > 2$. Then there exists a rational function $R_k(z)$, $k = 2m+1,\ldots$, which have at least $\frac{k}{2} - m$ poles lying on the rays $\rho e^{it_j}$, $1 < \rho < +\infty$, $j = 1,2,\ldots,m$, and the remaining poles at $\infty$, such that

$$\max_{|z|\leq 1} |f(z) - R_k(z)| \leq M_R k^A e^{-Dk^{\frac{1}{c}}}, \tag{2.23}$$

where $M_k$ is bounded as $R \to +\infty$ and

$$A = \max\left(\frac{2}{c}, \frac{2a+1}{2c}\right)$$

$$D = \frac{c}{2e}(abB_1)^{-\frac{1}{c}}, \qquad B_1 = \max\left(\frac{1}{R}, \frac{\pi}{2\beta}\right)$$

$$\beta = \min_j (t_{j+1} - t_j)/2.$$

REMARK. In our filter realization we take $f(z) \in H_m^R(a,b_2,4)$, where

$b_2 > \frac{24}{e^4}$ and $a$ is some constant depending on $b_2$. $f(z)$ has real coefficients and $f(e^{it})$-even function of $t$.

The order of best approximation by rational functions with coefficients equal to one in Bers space has been studied in [38]. It is connected with the approximation by electrostatic fields due to electrons.

Let $G$ be a simply connected domain with its boundary $\Gamma$ and $A_q(G)$, $1 < q < +\infty$ be a Bers space, it means that every function $f(z) \in A_q(G)$ is analytic in $G$ and has the bounded norm:

$$\|f\|_q := \iint_G |f(z)|\lambda_G^{2-q}(z)dxdy < +\infty \tag{2.24}$$

where $\lambda_G(z)$-Poincaré metric:

$$\lambda_G(z) = \frac{|\varphi'(z)|}{1-|\varphi(z)|^2} \tag{2.25}$$

and $w = \varphi(z)$ is a function which conformally maps $G$ onto $|w| < 1$ with $\varphi(z_0) = 0$, $\varphi'(z_0) > 0$, $z_0$ is some fixed point in $G$. Denote by $z = \psi(w)$ its inverse function.

This space is a generalization of the famous Hardy space $H(|z|<1)$. We are interested in the order of best approximation by rational functions with coefficients equal to one.

Let

$$E_{n,q}(S_\alpha) = \sup_{f(z)\in S_\alpha} \inf_{\substack{z_k\in\Gamma \\ 1\le k\le n}} \|f(z) - S_n(z)\|_q, \tag{2.24}$$

where

$$S_n(z) = \sum_{k=1}^{n} \frac{1}{z-z_k}, \quad z_k \in \Gamma, \quad k = 1,2,\dots,n \tag{2.25}$$

$$S_\alpha = \{f(z) \in A_q(D) \mid \|f\|_q \le 1, \ \omega(\delta,f) = 0(\delta^\alpha), \ 0 < \alpha \le 1 \tag{2.26}$$

and

$$\omega(\delta;f) = \sup_{|h|<\delta} \|f[\psi(w+h)] - f[\psi(w)]\|_q$$

$$:= \sup_{|h|<\delta} \iint_{|w|<1} |f[\psi(w+h)] - f[\psi(w)]|\,|\psi'(w)|^q(1- w^2)^{q-2}dudv.$$

There are some papers investigating the problem of completeness of polynomials in Bers space [39], [40], [41], [42], but for approximation by the rational functions of type (2.25) we knew only two papers [43], [44]. In 1973, Chui [43] proved: if the domain $G$ is bounded by a closed rectifiable Jordan curve and $q > 2$, then for every $A_q(G)$,

$$\lim_{n\to+\infty} \inf_{\substack{z_k \in C \\ k=1,2,\ldots,n}} \|f(z) - S_n(z)\| = 0 \tag{2.28}$$

is valid, where $S_n(z)$ is defined by (2.25). In 1972 Neuman stated: for $G = \{|z|<1\}$ and any $z_k$, $|z_k| = 1$, $k = 1,2,\ldots,n$

$$\|S_n\|_2 \geq \frac{\pi}{18} .$$

Hence, when $1 < q \leq 2$, the result (2.28) is not always valid for every function $f(z) \in A_q(|z|<1)$.

Chui and Shen [38] obtained the order of approximation by rational functions of type (2.25) in the case of $2 < q < +\infty$ and pointed out that their result in general is exact.

THEOREM 1 (Chui and Shen [38]). Let $2 < q < +\infty$, the boundary $\Gamma$ of $G$ be a Jordan curve of class $C^{2+\varepsilon}$ for some $\varepsilon > 0$, that is $\Gamma$ has a parametric representation $z = z(t)$, $z'(t) \neq 0$ on $\Gamma$, $z''(t) \in \mathrm{Lip}\,\varepsilon$. Then

1. $E_{n,q}(S_\alpha) = 0\left(\dfrac{1}{n^{q-2}}\right), \quad 2 < q \leq \dfrac{-\alpha+\sqrt{\alpha^2+12\alpha+16}}{2}$,

2. $E_{n,q}(S_\alpha) = 0\left(\dfrac{1}{n^{\frac{\alpha}{q+\alpha+2}}}\right), \quad q > \dfrac{-\alpha+\sqrt{\alpha^2+12\alpha+16}}{2}$.

Furthermore, the order of approximation in 1 is best possible.

Let $k$ be a positive integer and $S^k$, $0 < \alpha \leq 1$ be the class of functions $f \in A_q(G)$ such that $\|f\|_q \leq 1$, $f^1,\ldots,f^{(k)} \in A_q(G)$ and $\omega(\delta;f^{(k)}) = 0(\sigma^\alpha)$. For this class of functions we derived better order of approximation.

THEOREM 2 (Chui and Shen [38]). Let $2 < q < +\infty$, $k \geq 1$ and the boundary $\Gamma$ of $G$ be of class $C^{k+1+\varepsilon}$ for some $\varepsilon > 0$; that is $\Gamma$ has a parametric representation $z = z(t)$ which is $k+1$ times continuously differentiable with $z'(t) \neq 0$ and $z^{(k+1)}(t) \in \mathrm{Lip}\,\varepsilon$. Then

$$E_{n,q}(S_\alpha^k) = \begin{cases} O\left(\dfrac{1}{n^{q-2}}\right), & 2 < q \le \dfrac{\sqrt{(k+\alpha)^2+12(k+\alpha)+16}-(k+\alpha)}{2} \\ O\left(\dfrac{1}{n^{\frac{k+\alpha}{k+\alpha+q+2}}}\right), & q > \dfrac{\sqrt{(k+\alpha)^2+12(k+\alpha)+16}-(k+\alpha)}{2} \end{cases}$$

## §3. THE EXPANSION AND ESTIMATION OF REMAINDER BY MEANS OF RATIONAL FUNCTIONS

Let $\{\alpha_k\}$, $|\alpha_k| < 1$, $k = 0,1,\ldots$ and $\{\beta_k\}$, $|\beta_k| > 1$, $k = 1,2,\ldots$ be sequences of complex numbers. Let

$$(3.1) \quad \varphi_0(z) = \sqrt{\frac{1-|\alpha_0|^2}{2\pi}}\,\frac{1}{1-\bar{\alpha}_0 z}, \quad \varphi_n(z) = \sqrt{\frac{1-|\alpha_n|^2}{2\pi}}\,\frac{1}{1-\bar{\alpha}_n z}\prod_{k=0}^{n-1}\frac{z-\alpha_k}{1-\bar{\alpha}_k z}$$

$$(3.2) \quad \psi_0(z) = \sqrt{\frac{|\beta_1|^2-1}{2\pi}}\,\frac{1}{1-\bar{\beta}_1 z}, \quad \psi_m(z) = \sqrt{\frac{|\beta_m|^2-1}{2\pi}}\,\frac{1}{1-\bar{\beta}_m z}\prod_{k=0}^{m-1}\frac{z-\beta_k}{1-\bar{\beta}_k z}$$

$$n,m = 1,2,\ldots .$$

They are orthonormal systems on $|z| = 1$ that is if

$$(3.3) \quad \Phi_n(z) = \begin{cases} \varphi_n(z), & n \ge 0 \\ \psi_n(z), & n < 0, \end{cases}$$

then

$$(3.4) \quad \int_{|z|=1} \Phi_i(z)\overline{\Phi_j(z)}\,|dz| = \begin{cases} 1, & i = j \\ 0, & i \ne j . \end{cases}$$

Every function $f(z) \in L(|z|=1)$ can set a Fourier series about this system:

$$(3.5) \quad f(z) \sim \sum_{k=-\infty}^{+\infty} c_k\Phi_k(z) = \sum_{k=0}^{+\infty} c_k\varphi_k(z) + \sum_{k=1}^{+\infty} d_k\psi_k(z),$$

where

$$(3.6) \qquad c_k = \int_{|z|=1} f(z)\overline{\Phi_k(z)}|dz|, \qquad k = 0,+1,+2,\ldots .$$

The systems (3.3) has been first constructed by Malmquist and systematically by Walsh [45] and other authors. Under some conditions on the function $f(z)$ and distribution on $\{\alpha_k\}$ and $\{\beta_k\}$ Dzarbasjan obtained a expansion theorem on this system [46]. Ketbalian investigated the problem of expansion on systems of functions $\{\operatorname{Re} \varphi_n(z), \operatorname{Im} \varphi_n(z)\}$ for the function $f(z) \in L^p(|z|=1)$, $p \geq 1$ in the case of $\alpha_k = \frac{1}{\overline{\beta}_k}$, $k = 1,2,\ldots$ [47]. In these investigations the $\{\alpha_k\}$ and $\{\beta_k\}$ satisfy the conditions:

$$(3.7) \qquad \sum_{n=1}^{\infty} (1 - |\alpha_n|) = +\infty$$

$$(3.8) \qquad \sum_{n=1}^{\infty} \left(1 - \frac{1}{|\beta_n|}\right) = +\infty.$$

Let us consider the domain $G$ bounded by a closed rectifiable Jordan curve $\Gamma$ and rational function system $\{M_n^{(s)}(z)\}$ as defined by the sum of the principal parts of $\varphi_n[\Phi(z)]\Phi'(z)^s$ at their poles $b_k$:

$$(3.9) \qquad \Phi(b_k) = \frac{1}{\overline{\alpha}_k}, \qquad k = 0,1,2,\ldots,n$$

where $\Phi(z)$ is a mapping function (see §1). Dzarbasjan [48], [49] investigated the expansion on $M_n^{(s)}(z)$, $0 \leq s \leq 1$ for some classes of functions under the condition on $\Gamma$:

$$(3.10) \qquad \int_0^{2\pi} |\Psi'(e^{i\theta})|^{2(1-s)} d\theta < +\infty, \qquad 0 \leq s \leq 1.$$

Tymarkin studied the expansion on $\{M_n^{(\frac{1}{2})}(z)\}$ for some class of functions without any restriction on the boundary $\Gamma$ of $G$ [51]. Loukaski obtained the expansion on $\{M_n^{(1)}(z)\}$ for class of functions $E_p$, $p > 1$ under the condition [52]

$$\overline{\lim_{r\to k+0}} \int_0^{2\pi} |\psi'(re^{i\alpha})|^k d\alpha < +\infty, \qquad k > 1.$$

Shen [29], [53] generalized these results by the following theorems.

THEOREM 1 (Shen [29], [53]). Under the condition

$$(3.10) \qquad \sum_{k=1}^{n} \left(1 - \frac{1}{|\Phi(b_k)|}\right) = +\infty$$

where $\{b_k\}$ is the sequence in $G_\infty$ and $f(z) \in E_p(G)$, $p > 1$, we have the expansion

$$(3.11) \qquad f(z) = \sum_{k=0}^{\infty} a_k M_k^{(\frac{1}{p})}(z), \qquad z \in G,$$

which converges in the interior of $G$ uniformly, where

$$(3.12) \quad a_k = a_k(f) = \frac{1}{i} \int_{|\tau|=1} f[\Psi(\tau)]\Psi'(\tau)^{\frac{1}{p}} \overline{\frac{1}{\tau}\varphi_k\left(\frac{1}{\tau}\right)} d\tau, \qquad k = 0,1,\ldots,$$

are called Fourier coefficients of $f(z)$ on $\{M_n^{(\frac{1}{p})}(z)\}$.

For the case of $p \geq 1$, we have

THEOREM 2 (Shen [29] [53]). Under the condition (3.11) and $f(z) \in E_p(G)$, $p \geq 1$, we have the expansion

$$(3.13) \quad f(z) = \lim_{R\to 1+0} \sum_{k=0}^{\infty} \frac{1}{i} \int_{|\tau|=1} f[\Psi(\tau)]\Psi'(\tau)^{\frac{1}{p}} \overline{\frac{1}{R\tau}\varphi_k\left(\frac{1}{R\tau}\right)} d\tau \cdot M_k^{(\frac{1}{p})}(z), \; z \in G,$$

which converges in the interior of $G$ uniformly and it is called generalized Abel's summation.

THEOREM 3 (Shen [29] [53]). Let $G \in K_q$, $q > 1$. If we define the operator

$$(3.14) \qquad (TR)(z) = \frac{1}{2\pi i} \int_\Gamma \frac{R[\Phi(\xi)]\Phi'(\xi)^{\frac{1}{p}} d\xi}{\xi - z}, \qquad \frac{1}{p} + \frac{1}{q} = 1$$

where $R(w)$ is a rational function with its poles in $|w| > 1$ and its norm is defined as $\|R(w)\|_{L_p(|w|=1)}$ (This space is called $S_{|w|<1}$), we have

1. $TR$ is a bounded operator in space $S_{|w|<1}$ and its image is in $E_p(G)$. Moreover, it can be expanded to $H^p(|w|<1)$ without changing its norm.

2. $(T\varphi_n(z))(z) = M_n^{(\frac{1}{p})}(z)$.

3. If $g(w) \in H_p(|w|<1)$, then from $(Tg)(z) = 0$ it follows that $g(w) \equiv 0$.

4. There exists a unique inverse operator in this space, i.e., if $F(z) \in T(H_p)$, then the inverse operator has the form

$$(3.15) \qquad (NF)(w) = \frac{1}{2\pi i}\int_{|\tau|=1} \frac{F[\Psi(\tau)]\Psi'(\tau)^{\frac{1}{p}}d\tau}{\tau-w}, \qquad |w|<1.$$

There are some papers contributed to the order of approximation by the partial sum of Fourier expansion of a function on the system $\{M_n^{(1/p)}(z)\}$ under some condition on the boundary $\Gamma$ of domain $G$.

Let $f(z) \in L^p(|z|=1)$, $p > 1$ and $S_{n,m}(z)$ be a partial sum in (3.5) of $f(z)$:

$$(3.16) \qquad S_{n,m}(z) = \sum_{k=0}^{n} c_k\varphi_k(z) + \sum_{k=1}^{m} d_k\psi_k(z).$$

Then Loukaski [54] obtained

$$(3.17) \qquad \|S_{n,m}(z)\|_{L^p(|z|=1)} \le C_p\|f(z)\|_{L^p(|z|=1)}.$$

where $C_p$ is a constant. Therefore, it follows

$$(3.18) \qquad \|f(z) - S_{n,m}(z)\|_{L^p(|z|=1)} \le (C_p+1)\|f(z) - B_{n,m}(z)\|_{L^p(|z|=1)}.$$

where $B_{n,m}(z)$ is the best approximant of $f(z)$ by rational functions with its poles at $\alpha_i$, $i = 1,2,\dots,n$ and $\beta_j$, $j = 1,2,\dots,m$. Ketbalian obtained the similar result only in the case of $\beta_k = \frac{1}{\bar{\alpha}_k}$, $k = 0,1,\dots$ [47].

Coqialian [19] has obtained the estimation in space $C(|z|=1)$:

$$(3.19) \qquad \|S_{n,m}(z)\|_{C(|z|=1)} \le (\pi + \ln r_{n,m})\|f(z)\|_{C(|z|=1)},$$

where

$$(3.20) \qquad r_{n,m} = \sum_{k=0}^{n}(1-|a_k|)^{-1} + \sum_{k=1}^{m}\left(1-\frac{1}{|\beta_k|}\right)^{-1}.$$

Hence

$$(3.21) \qquad \|f(z) - S_{n,m}(z)\|_{C(|z|=1)} \le (\pi + 1 + \ln r_{n,m})\|f(z) - B_{n,m}(z)\|_{C(|z|=1)}.$$

Su [4] obtained the same estimation in $L'(|z|=1)$.

Coqialian [19] considered the approximation in Alper's domain, i.e., $j_1$ domain (see §1) and obtained some results by applying the Mergelian and

Dzarhasjan's estimation [12], but his result is not exact. Shen [55] considered the approximation in a domain $G$ the boundary $\Gamma$ of which satisfies the condition (2.15) and obtained the order of approximation by the partial sum of expansion on $\left\{M_n^{(\frac{1}{p})}(z)\right\}$ for class of functions $E_p(G)$, $p > 1$. But the better result was given in Shen's papers [29] [53].

THEOREM 4 (Shen [29] [53]). Let $G \in K_q$, $q > 1$, $f(z) \in E_p(G)$, $\frac{1}{p}+\frac{1}{q}=1$ and

$$(3.22) \qquad S_n(z) = \sum_{k=0}^{n} a_k M_p^{(\frac{1}{p})}(z),$$

be the partial sum of Fourier expansion of $f(z)$, where $a_k$, $k = 1,2,\ldots,n$, are defined by (3.2), then there is a constant $C$ such that

$$(3.23) \qquad \|f(z) - S_n(z)\|_{L_p(\Gamma)} \le C\|F(w) - B_n(w)\|_{L_p(|w|=1)},$$

where $F(w) = (Nf)(w)$ (see (3.15)), and $B_n(w)$ is the best approximant of $F(w)$ by rational functions with its poles at $\Phi(b_k)$, $k = 0,1,\ldots,n$.

Applying the result [11] to this theorem the following theorem follows naturally.

THEOREM 5 (Shen [29] [53]). If $G \in K_q$, $q > 1$, then for any $f(z) \in E_p(G)$, $\frac{1}{p}+\frac{1}{q} = 1$, we have

$$(3.24) \qquad \|f(z) - S_n(z)\|_{L_p(\Gamma)} \le C\left[\omega(\varepsilon_n(\alpha);F) + q^{\frac{1}{\varepsilon_n(\alpha)}}\right],$$

where $\varepsilon_n(\alpha)$ is defined by (2.1), $F(w) = (Nf)(w)$ (see (3.15)) and $q$ is a absolute constant, $0 < q < 1$.

Moreover, Shen [21] obtained the estimation in space $C(\Gamma)$ too.

THEOREM (Shen [21]). Let $G$ be a $j_\lambda$ domain (see §1), $0 \le \lambda \le 1$, $f(z)$ be analytic in $G$ and for some integer $k \ge 0$, $f^{(k)}(z)$ be continuous on $\overline{G} = G + \Gamma$, then

$$(3.25) \quad \|f(z) - S_n(z)\|_{C(\Gamma)}$$

$$\le C\left[\ln \sum_{k=0}^{n} (1-|\alpha_k|)^{-1}\right] \cdot \left[\ln \sum_{k=0}^{n} (1-|\alpha_k|)^{-1} + \ln \sum_{k=0}^{n} (1-|\alpha_k|)\right]^{1-\lambda} R_n(f;\alpha_k)$$

$$\le C\left[\ln \sum_{k=0}^{n} (1-|\alpha_k|)^{-1}\right] \cdot \left[\ln \sum_{k=0}^{n} (1-|\alpha_k|)^{-1} + \ln \sum_{k=0}^{n} (1-|\alpha_k|)\right]^{1-\lambda} \cdot$$

$$\left[(\varepsilon_n(\alpha))^k \omega(\varepsilon_n(\alpha);f^{(k)}) + q^{\frac{1}{\varepsilon_n(\alpha)}}\right],$$

where $\alpha_k = \dfrac{1}{\Phi(b_k)}$, $k = 0,1,\ldots,n$, $R_n(f;\alpha_k)$ is the best approximation of $f(z)$ rational functions with its poles at $\Phi(b_k)$, $k = 0,1,\ldots,n$, $\varepsilon_n(\alpha)$ is defined by (2.1) and $q$ is a absolute constant, $0 < q < 1$.

## §4. THE INCOMPLETENESS, BASIS AND MOMENT PROBLEM OF SOME SYSTEMS OF FUNCTIONS

There are a lot of papers investigating the basis of rational functions and the problem of interpolation in $H^p$ spaces in the unit circle as well as over a half-plane (see, for example, [56], [57], [58], [59] and [60]).

In 1977-1978 M.M. Dzarbasjan obtained the efficient solution for multiple interpolation in $H^p$ spaces, $1 < p < \infty$ over a half-plane [60] [61]. In 1984 Wu solved this problem in $H^p$ spaces $0 < p < 1$.

THEOREM 1 (Wu [62]). Suppose the sequence $\{\lambda_j\}$, $\operatorname{Im} \lambda_j > 0$, $j = 1,2,\ldots,$ satisfies the condition:

$$(4.1) \qquad \inf_k \prod_{\lambda_j \ne \lambda_k} \left|\frac{\lambda_j-\lambda_k}{\bar{\lambda}_j-\lambda_k}\right| \ge \delta > 0$$

and denote by $S_k$ the number of appearance of $\lambda_k$ in $\{\lambda_1,\lambda_2,\ldots,\lambda_k\}$ and by $p_k$ the number of appearance of $\lambda_k$ in $\{\lambda_j\}$:

$$(4.2) \qquad \sup\{S_k\} = \sup\{p_k\} = P < +\infty.$$

Let $\{w_j\}$ be a sequence of complex numbers satisfying:

$$(4.3) \quad \|w\|_j^{\{\lambda_k\}} := \left\{\sum_{j=1}^{\infty} (\operatorname{Im} \lambda_j)^{p(s_j-1)+1} |w_j|^p\right\}^{\frac{1}{p}} < +\infty, \qquad 0 < p < \infty.$$

Then by taking $n$, $\frac{1}{n+1} < p < +\infty$, the series

$$(4.4) \qquad \sum_{k=1}^{\infty} w_k \Omega_{k,n}(z) = f(z) \in H^p(\operatorname{Im} z > 0)$$

uniformly converges in $\operatorname{Im} z > 0$ and satisfies:

1. $f^{(s_j-1)}(\lambda_j) = w_j, \quad j = 1,2,\ldots,$

2. $\|f\|_p \le C\|w_j\|^{\{\lambda_k\}},$

where $C$ is a constant depending on $P$, $\delta$, $n$ and $p$ and the functions $\Omega_{k,n}(z)$ are defined as follows:

Let

$$(4.5) \qquad B(z) = \prod_{j=1}^{\infty} \frac{z-\lambda_j}{z-\bar{\lambda}_j} \frac{|1+\lambda_j^2|}{1+\lambda_j^2}$$

and

$$\frac{(z-\bar{\lambda}_k)^n (z-\lambda_k)^{p_k}}{B(z)} = \sum_{\nu=0}^{\infty} \alpha_{\nu,n}(\lambda_k)(z-\lambda_k)^{\nu}$$

then define $\Omega_{k,n}(z)$ as:

$$(4.6) \quad \Omega_{k,n}(z) = \frac{(z-\lambda_k)^{s_k-1}}{(s_k-1)!} \frac{B(z)}{(z-\bar{\lambda}_k)^n (z-\lambda_k)^{p_k}} \sum_{\nu=0}^{p_k-s_k} \alpha_{\nu,n}(\lambda_k)(z-\lambda_k)^{\nu},$$

$$k = 1,2,\ldots .$$

THEOREM 2 (Wu [62]). Suppose the sequence $\{\lambda_j\}$, $\operatorname{Im} \lambda_j > 0$, $j = 1,2,\ldots$ satisfies the (4.1) and (4.2), then for $\frac{1}{n+1} < p < \infty$, $\{\Omega_{k,n}(z)\}$ is the basis of the closure of $\{\Omega_{k,n}(z)\}$.

DEFINITION (see [60]). Denote by $H^p_{\pm}\{\lambda_j\}$, $p > 0$, the class of functions $f(z)$ satisfying the following conditions:

1. $f(z) \in H^p(\operatorname{Im} z > 0)$,

2. $f(z) = B(z)\tilde{f}(z)$, $\tilde{f}(z) \in H^p(\operatorname{Im} z < 0)$,

where $B(z)$ is defined by (4.5).

3. $f(z) = B(x)\tilde{f}(x)$ almost everywhere on $(-\infty,+\infty)$.

THEOREM 3 (Wu [62]). If $0 < p < 1$, then $H^p_1\{\lambda_j\}$ has not interpolation basis, i.e. for any $\{g_k(z)\}$, $g_k(z) \in H^p_{\pm}\{\lambda_j\}$:

$$(4.7) \qquad g_k^{(s_j-1)}(\lambda_j) = \delta_{j,k} = \begin{cases} 0, & j \neq k \\ 1, & j = k, \end{cases}$$

there exists a $f_0(z) \in H^p_{\pm}\{\lambda_j\}$:

$$f_0(z) \neq \sum_{j=0}^{\infty} f_0^{(s_j-1)}(\lambda_j) g_j(z).$$

As we know that under the condition (4.2) and

$$(4.8) \qquad \sum_{j=1}^{\infty} \frac{\operatorname{Im} \lambda_j}{1+|\lambda_j|^2} < +\infty,$$

if $1 < p < +\infty$, the necessary and sufficient condition for that $\left\{\frac{(s_k-1)!}{(z-\overline{\lambda_k})^{s_k}}\right\}$ is the basis in $H^p_{\pm}\{\lambda_j\}$ is the condition (4.1).

For the case of $p = +\infty$ Shen [63] got the efficient solution for the multiple interpolation in $H^\infty$ over a half-plane.

THEOREM (Shen [63]). Suppose $\{\lambda_j\}$ satisfies the conditions (4.1) and (4.2), and $\{w_j\}$ satisfies the condition:

$$(4.9) \qquad \|w_k\|_\infty^{\{\lambda_j\}} := \sup (\operatorname{Im} \lambda_k)^{s_k-1} |w_k| < +\infty,$$

then there exists a function $f_0(z) \in H^\infty$ $(\operatorname{Im} z > 0)$ satisfying the following conditions:

1. $f_0^{(s_k-1)}(\lambda_k) = w_k$, $k = 1,2,\ldots,$
2. $\|f_0\|_\infty \leq C\|w_k\|_\infty^{\{\lambda_j\}}$,
3. $f_0(z) = \sum_{k=1}^{\infty} w_k \hat{\Omega}_k(z)$,

which converges absolutely and uniformly in $\operatorname{Im} z > 0$ and the functions $\hat{\Omega}_k(z)$ are defined as follows:

1. Let $\{z_k\}$ be a sequence containing all different elements of $\{\lambda_j\}$:

$$\left|\frac{z_k - i}{z_k + i}\right| \le \left|\frac{z_{k+1} - i}{z_{k+1} + i}\right|, \qquad k = 1,2,\ldots$$

2. For arbitrary $\lambda_n$ we can find $z_j = \lambda_n$. Consider

$$G(z,\lambda_n) = \exp\left[-i \sum_{k\ge j} \frac{4 \operatorname{Im} z_k}{1+|z_k|^2+2\operatorname{Im} z_k} \cdot \frac{1+\bar{z}_k z}{z-z_k}\right]$$

and

$$G_1(z,\lambda_n) = \left(\frac{i \operatorname{Im} \lambda_n}{\lambda_{n+i}} \cdot \frac{z+i}{z-\bar{\lambda}_n}\right)^{p_{n+1}} G(z,\lambda_n)$$

3. Let $z = i\,\frac{1+w}{1-w}$, then $\lambda_n$ corresponds some $a_n, n = 1,2,\ldots$ . Let

$$\tau_n(z) = \frac{(z-\lambda_n)^{p_n}}{B(z)G_1(z,\lambda_n)} = \sum_{\nu=0}^{\infty} \alpha_\nu(a_n)(w-a_n)^\nu,$$

and

$$\hat{\Omega}_n(z) = \frac{(z-\lambda_n)^{s_n-1}}{(s_n-1)!} \frac{B(z)G_1(z,\lambda_n)}{(z-\lambda_n)^{p_n}} \sum_{\nu=0}^{p_n-s_n} \alpha_\nu(a_n)\left[\frac{2i(z-\lambda_n)}{(\lambda_{n+i})(z+i)}\right]^\nu .$$

By the way we would like to indicate that in 1981 Martirosian obtained the efficient solution for multiple interpolation problem in $H^\infty$ space in the unit circle [64].

Let $\{a_k\}$, $|a_k| < 1$, $k = 1,2,\ldots$ be a sequence the elements of which can be coincided with each other. We denote by $s_n$ the number of appearance of $a_n$ in $\{a_1,a_2,\ldots,a_n\}$ and by $p_n$ the number of appearance of $a_n$ in $\{a_j\}$. Suppose

$$\sup\{s_n\} = \sup\{p_n\} = p < +\infty \tag{4.10}$$

$$\sum_{k=1}^{\infty} (1-|a_k|) < +\infty. \tag{4.11}$$

Under these conditions the Blaschke product

$$B(w) = \prod_{k=1}^{\infty} \frac{a_k - w}{1-\bar{a}_k w} \frac{|a_k|}{a_k} \tag{4.12}$$

exists. Consider

$$\frac{(w-a_k)^{p_k}}{B(w)} = \sum_{\nu=0}^{\infty} \alpha_\nu(a_k)(w-a_k)^\nu$$

and

$$(4.13)\quad \Omega_k(w) = \frac{(w-a_k)^{s_k-1}}{(s_k-1)!}\;\frac{B(w)}{(w-a_k)^{p_k}}\sum_{\nu=0}^{p_k-s_k}\alpha_\nu(a_k)(w-a_k)^\nu,\quad k=1,2,\ldots$$

$$(4.14)\qquad \gamma_k(w) = \frac{(s_k-1)!\,w^{s_k-1}}{(1-\bar{a}_k w)^{s_k}},\qquad k=1,2,\ldots .$$

It is well known that $\{\gamma_k(w)\}$ and $\{\Omega_k(w)\}$ are biorthonormal systems on $|w| = 1$ [59], i.e.

$$(4.15)\quad \frac{1}{2\pi}\int_{|w|=1}\gamma_k(w)\overline{\Omega_n(w)}\,|dw| = \frac{1}{2\pi}\int_{|w|=1}\overline{\gamma_k(w)}\Omega_n(w)\,|dw| = \delta_{kn}.$$

We can pose the moment problem for $\{\Omega_k(w)\}$: For any given sequence of complex numbers $\{g_k\}$, can we find out a function $f(z) \in H^p(|w|<1)$, $1 \le p \le +\infty$ such that

$$(4.16)\qquad \frac{1}{2\pi}\int_{|\tau|=1} f(\tau)\overline{\Omega_k(\tau)}\,|d\tau| = g_k,\qquad k=1,2,\ldots\ ?$$

If the solution of moment problem exists, is it unique? If it is not unique, when is it?

DEFINITION. The sequence $\{a_k\}$ is said to be belonging to $\Delta(P,\delta)$, if it satisfies condition (4.10) and

$$(4.17)\qquad \inf_k \prod_{a_j\ne a_k}\left|\frac{a_j-a_k}{1-\bar{a}_j a_k}\right| \ge \delta > 0.$$

THEOREM 1 (Shen [65]). If $\{a_k\} \in \Delta(P,\delta)$, then the necessary and sufficient condition for the existence of solution to the moment problem (4.16) in $H^p(|w|<1)$, $1 < p < \infty$ is:

$$(4.18)\qquad \|g_k\|_p^{\{a_j\}} := \left\{\sum_{k=1}^{\infty}(1-|a_k|^2)^{-ps_k+1}|g_k|^p\right\}^{\frac{1}{p}} < +\infty$$

and the solution can be expressed as follows:

$$(4.19) \qquad f_1(w) = \sum_{k=1}^{\infty} g_k\gamma_k(w), \qquad f_1(w) \in H^p(|w|<1),$$

the series (4.19) converges in $L_p(|w|=1)$ strongly and converges in the interior of $|w| < 1$ uniformly. Besides

$$(4.20) \qquad \|f_1(w)\|_{L_p(|w|=1)} \leq C\|g_k\|^{\{a_j\}}.$$

THEOREM 2 (Shen [65]). Let $\{a_k\} \in \Delta(P,\delta)$ and $\{g_k\}$ satisfy the condition (4.18), then all the solutions $f(w)$ to the moment problem (4.16) can be expressed as follows:

$$f(w) = f_1(w) + f_2(w)$$

where

$$f_1(w) = \sum_{k=1}^{\infty} g_k\gamma_k(w), \qquad |w| < 1,$$

$f_1(w) \in \lambda_p(a_k)$ [66], i.e.

1. $f(w) \in H^p(|w|=1)$.

2. $f(w) = B(w)\tilde{f}(w)$, $\tilde{f}(w) \in H^p\ (|w|>1)$, $\tilde{f}(\infty) = 0$, where $B(w)$ is defined by (4.12).

3. $f(w) = B(w)\tilde{f}(w)$, almost everywhere on $|w| = 1$, and

$$f_2(w) = B(w)\tilde{f}_2(w), \quad \tilde{f}_2(w) \in H^p(|w|<1),$$

where $B(w)$ is the Blaschke product (4.12).

THEOREM 3 (Shen [65]). Suppose $\{a_k\}$ satisfies the condition (4.10) and (4.11). Then the necessary and sufficient condition for solving the problem (4.16) in $H^p\ (|w|<1)$, $p > 1$ for any arbitrary sequence $\{g_k\}$ of complex numbers is $\{a_k\} \in \Delta(P,\delta)$.

Now for any given function $f(w) \in H^p(|w|<1)$, $p > 1$, define $\{g_k\}$ according to (4.16), then introduce an operator:

$$(4.21) \qquad M_p : f \in H^p(|w|<1) \to \left\{(1-|a_k|)^{-s_k+\frac{1}{p}} g_k\right\}.$$

THEOREM 4 (Shen [65]). Suppose $\{a_k\}$ satisfies the conditions (4.10) and (4.11). Then the necessary and sufficient condition for $M_p(H^p) = \ell^p$, $p > 1$ is $\{a_k\} \in \Delta(P,\delta)$.

Some of above established results can be generalized to the domain of complex plane.

Let $G$ be a domain bounded by a closed rectifiable Jordan curve $\Gamma$ in the complex plane, $G_\infty$ be a complement of $\overline{G}$. In 1981 Shen [30] proved that the necessary and sufficient condition for the completeness of $\left\{\frac{1}{(z-\alpha_i)^{s_i}}\right\}$, $\alpha_i \in G_\infty$ in $E_p(G)$, $\frac{1}{q}+\frac{1}{p} = 1$, $G \in K_q$, $q > 1$ is

$$(4.22) \qquad \sum_{k=1}^{\infty} (1-|a_k|) = +\infty, \qquad a_k = \overline{\Phi(\alpha_k)^{-1}},$$

where $\Phi(z)$ is a mapping function (see §1).

If

$$(4.23) \qquad \sum_{k=1}^{\infty} (1-|a_k|) < +\infty , \qquad a_k = \overline{\Phi(\alpha_k)^{-1}},$$

then the closure of $\left\{\frac{1}{(z-\alpha_i)^{s_i}}\right\}$ is a subspace which we denote by $R_p(G;\alpha_i)$ in $E_p(G)$. We are interested in the characteristic properties of $R_p(G;\alpha_i)$.

Let

$$(4.24) \quad \gamma_k(w) = \frac{(s_k-1)!w^{s_k-1}}{(1-\overline{a}_k w)^{s_k}}, \qquad a_k = \overline{\Phi(\alpha_k)^{-1}}, \qquad k = 1,2,\ldots,$$

and $m_k^{(\frac{1}{p})}(z)$ be the principal part of $\gamma_k[\Phi(z)]\Phi'(z)^{\frac{1}{p}}$ at $z = \alpha_k$, $1 < p < \infty$. Obviously, the closure of $\left\{m^{(\frac{1}{p})}(z)\right\}$ is the $R_p(G;\alpha_i)$ too.

For $\Omega_n(w)$ (see (4.13)) we construct

$$(4.25) \qquad \rho_k^{(\frac{1}{p})}(z) = \frac{\Phi'(z)}{\Phi(z)}\,\Omega_k\left(\frac{1}{\Phi(z)}\right), \qquad k = 1,2,\ldots .$$

It can be proved that

$$(4.26) \qquad \frac{1}{2\pi i}\int_P m_n^{(\frac{1}{p})}(\zeta)\rho_m^{(\frac{1}{p})}(\zeta)d\zeta = \delta_{n,m},$$

(for special case $p = 2$ see [67], for different $\{\alpha_i\}$ see [68], also see [69]).

DEFINITION. The function $f(z)$ is said to belong to $\lambda_p(G;\alpha_i)$, if $f(z) \in E_p(G)$, $p > 1$ and satisfies

$$(4.27)\qquad \frac{1}{2\pi i}\int_{|t|=1}\frac{f[\Psi(t)]\Psi'(t)^{\frac{1}{p}}}{tB(t)}\gamma(\frac{1}{t};z)dt \equiv 0, \qquad z \in G,$$

where $B(t)$ is the Blaschke product (see (4.12)) and $\gamma(w;z)$ is defined by the following formula

$$(4.28)\qquad \gamma(w;z) = \frac{1}{2\pi i}\int_{|t|=1}\frac{B(t)\Psi'(t)^{\frac{1}{q}}}{\Psi(t)-z}\frac{dt}{1-tw} \in H_q(|w|<1), \quad \frac{1}{q}+\frac{1}{p}=1.$$

THEOREM 1 (Shen [70]). Let $\underline{G \in K_q \cap K_p}$, $\frac{1}{p}+\frac{1}{q}=1$, $p>1$, $\{\alpha_k\} \in \Delta(P,\delta)$, it means $\{a_i = \Phi(\alpha_i)^{-1}\} \in \Delta(P,\delta)$ (see (4.17)), then for any given $f(z) \in \lambda_p(G;\alpha_i)$ the series

$$(4.29)\qquad \sum_{k=1}^{\infty} \ell_k(f)m_k^{(\frac{1}{p})}(z) = f(z)$$

converges in $L_p(\Gamma)$ strongly, where $\ell_k(f)$ is defined by the following formula

$$(4.30)\qquad \ell_k(f) = \frac{1}{2\pi i}\int_{\Gamma} f(\zeta)\rho_k^{(\frac{1}{p})}(\zeta)d\zeta, \qquad k = 1,2,\ldots .$$

REMARK. By the conditions of the above theorem it is easy to see that

$$(4.31)\qquad R_p(G;\alpha_i) = \lambda_p(G;\alpha_i),$$

consequently, system $\left\{m^{(\frac{1}{p})}(z)\right\}$ is the basis in its closure $R_p(G;\alpha_i)$.

THEOREM 2 (Shen [70]). Under the condition of the above theorem the class $R_p(G;\alpha_i) = \lambda_p(G;\alpha_i)$ can be characterized as follows:

1. $f(z) \in E_p(G)$;

2. $f(z) = B[\Phi(z)F(z)$, $z \in G_\infty$, $F(z) \in E_p(G_\infty)$, $F(\infty) = 0$;

3. The boundary values of $f(z)$ from inside $\Gamma$ and outside $\Gamma$ are identical almost everywhere.

THEOREM 3 (Shen [71]). Let $G \in K_p$, $p>1$ and the sequence $\{\alpha_i\}$ in $G_\infty$ satisfy the conditions (4.10) and (4.23), then the system $\left\{\rho_k^{(\frac{1}{p})}(z)\right\}$ is not complete in $E_q^0(G)$, $\frac{1}{q}+\frac{1}{p}=1$, it means that there exists a function $g(z) \in E_q(G)$, $g(\infty) = 0$ such that

$$\liminf_{\substack{n\to\infty \\ b_i}} \|g(z) - \sum_{i=1}^{m} b_i \rho_i^{(\frac{1}{p})}(z)\|_{L_q(\Gamma)} > 0.$$

Thus the space $Q_p(G_\infty;\alpha_i)$ produced by the system $\left\{\rho_k^{(\frac{1}{p})}(z)\right\}$ is a real subspace in $E_q^0(G_\infty)$, so we are interested in its characteristic properties and other problems.

DEFINITION. The function $g(\zeta)$ is said to belong to $M_q(G_\infty;\alpha_i)$, if $g(\zeta) \in E_q^0(G_\infty)$ and satisfies

$$\frac{1}{2\pi i}\int_\Gamma g(z)\cdot Q(z;\zeta)dz \equiv 0, \qquad \zeta \in G_\infty, \tag{4.32}$$

where the function $Q(z;\zeta)$ is defined as follows:

$$Q(z;\zeta) = \frac{1}{2\pi i}\int_\Gamma \frac{B[\Phi(\eta)]\Phi'(\eta)^{\frac{1}{p}}d\eta}{(\eta-z)(\Phi(\zeta)-\Phi(\eta))}, \qquad z \in G, \qquad \zeta \in G_\infty. \tag{4.33}$$

THEOREM 4 (Shen [71]). The necessary and sufficient condition for $g(\zeta) \in M_q(G_\infty;\alpha_i)$ is that there exists a function $G(\tau) \in H^q(|\tau|<1)$ such that

$$G(\tau) = g[\Psi(\tau)]B(\tau)\Psi'(\tau)^{\frac{1}{q}}, \qquad |\tau| = 1 \tag{4.34}$$

holds almost everywhere, where $B(\tau)$ is defined by (4.12), and $\Psi(\tau)$ is a mapping function.

THEOREM 5 (Shen [71]). Let $G \in K_q \cap K_p$, $\frac{1}{p}+\frac{1}{q} = 1$, $q > 1$, $\{\alpha_i\} \in \Delta(P,\delta)$, then for any given $g(\zeta) \in M_q(G_\infty;\alpha_i)$ the series

$$\sum_{k=1}^{\infty} h_k(g)\rho_k^{(\frac{1}{p})}(\zeta) = g(\zeta) \tag{4.35}$$

converges in $L_q(\Gamma)$ strongly, where $h_k(g)$ is defined as follows

$$h_k(g) = \frac{1}{2\pi i}\int_\Gamma g(\zeta)m_k^{(\frac{1}{p})}(\zeta)d\zeta, \qquad k = 1,2,\ldots . \tag{4.36}$$

REMARK. From the above Theorem 4 it is easy to see

$$Q_q(G_\infty;\alpha_i) = M_q(G_\infty;\alpha_i). \tag{4.37}$$

Hence under the condition of Theorem 5 the system $\left\{\rho_k^{(\frac{1}{p})}(\zeta)\right\}$ is the basis in its closure $Q_q(G_\infty;\alpha_i)$.

For the system $\left\{\rho_k^{(\frac{1}{p})}(\zeta)\right\}$, $\zeta \in G_\infty$, we can pose the moment problem: for any given sequence $\{g_k\}$ of complex numbers, can we find out a function $f(z) \in E_p(G)$, $p > 1$ such that

$$\frac{1}{2\pi i}\int_\Gamma f(\zeta)\rho_k^{(\frac{1}{p})}(\zeta)d\zeta = g_k, \qquad k = 1,2,\ldots . \tag{4.38}$$

THEOREM 6 (Shen [72]). Let $G \in K_q \cap K_p$, $p > 1$, $\frac{1}{q}+\frac{1}{p} = 1$, $\{\alpha_i\} \in \Delta(P,\delta)$, then the necessary and sufficient condition for solving the moment problem (4.38) in $E_p(G)$ is

$$\|g_k\|^{\{\alpha_i\}} := \left\{\sum_{k=1}^\infty (1-|a_k|)^{-ps_k+1}|g_k|^p\right\}^{\frac{1}{p}} < +\infty, \quad a_k = \overline{\Phi(\alpha_k)^{-1}}, \tag{4.39}$$

$$k = 1,2,\ldots,$$

and in the case of existence of solution it can be expressed as follows:

$$f(z) = \sum_{k=1}^\infty g_k m_k^{(\frac{1}{p})}(z), \qquad z \in G, \tag{4.40}$$

the series (4.40) converges in $L_q(\Gamma)$ strongly and converges in the interior of $G$ uniformly. Besides $f(z) \in X_p(G;\alpha_i)$, i.e.

1. $f(z) \in E_p(G)$;
2. $f(z) = B[\Phi(z)]F(z)$, $z \in G_\infty$, $F(z) \in E_p^0(G_\infty)$;
3. the boundary values of $f(z)$ from two sides of $\Gamma$ are equal to each other on $\Gamma$ almost everywhere.

Furthermore,

$$\|f\|_{L_q(\Gamma)} \le C\|g_k\|^{\{\alpha_i\}}. \tag{4.41}$$

THEOREM 7 (Shen [72]). Under the condition on Theorem 6, if (4.39) is valid, then all the solution of moment problem (4.38) $f(z) \in E_p(G)$ can be expressed as follows:

$$f(z) = f_1(z) + f_2(z),$$

where $f_1(z) \in X_p(G;\alpha_i)$

$$f_1(z) = \sum_{k=1}^\infty g_k m_k^{(\frac{1}{p})}(z) = \sum_{k=1}^\infty \ell_k(f_1)m_k^{(\frac{1}{p})}(G), \qquad z \in G, \tag{4.42}$$

where $\ell_k(f_1)$ is defined by (4.30), and the series (4.42) converges in $L_p(\Gamma)$,

$f_2(\zeta) \in E_p(G)$ and function

$$F_2(w) = \frac{1}{2\pi i}\int_{|\tau|=1} \frac{f_2[\Psi(\tau)]\Psi'(\tau)^{\frac{1}{p}}}{\tau - w}\, d\tau \in H_p \quad (|w|<1)$$

possesses the expression

$$F_2(w) = G(w)\widetilde{F}_2(w), \qquad \widetilde{F}_2(w) \in H_p \quad (|w|<1).$$

THEOREM 8 (Shen [72]). Suppose $G \in K_q \cap K_p$, $q > 1$, $\frac{1}{q} + \frac{1}{p} = 1$. For any given $\{g_k\}$ satisfying (4.39), the necessary and sufficient condition for the moment problem (4.38) with a solution $f(z) \in E_p(G)$ is $\{\alpha_i\} \in \Delta(P,\delta)$.

Now for any given function $f(z) \in E_p(G)$, $p > 1$ define $g_k$ according to (4.38), then one can introduce an operator:

$$M_p : f \in E_p(G) \to \left\{(1-|\Phi(\alpha_k)|)^{-s_k+\frac{1}{p}} g_k\right\}.$$

From Theorem 6 we know that if $\{\alpha_i\} \in \Delta(P,\delta)$, then $M_p(E_p(G)) \subset \ell_p$. Generally by combining Theorem 6 and 8 we can directly obtain the following theorem.

THEOREM 9 (Shen [72]). Let $G \in K_q \cap K_p$, $q > 1$, $\frac{1}{q} + \frac{1}{p} = 1$ and sequence $\{\alpha_i\}$ satisfy the conditions (4.10) and (4.23), then the necessary and sufficient condition for $M_p(E_p(G)) = \ell_p$ is $\{\alpha_i\} \in \Delta(P,\delta)$.

## §5. THE CLOSURE OF $\{z^{\tau_n} \ln^j z\}$ ON UNBOUNDED CURVES AND DOMAINS

Up to the present, a great deal of work has been done in connection with the closure of the system $\{z^n\}$ or more generally the system of functions $\{z^{\tau_n}\}$ (where $\{\tau_n\}$ is a sequence satisfying certain conditions) on unbounded curves with respect to a certain weight function (see for example Saginjan [73], [74], Fuchs [75], Dzarbasjan [76], [77], Mandelbrojt [78], Dzarbasjan and Hacatrjan [79], Leont'ev [80] and Yu [81]). In their papers the following special cases were considered:

1. The curves at which the approximation is considered is real axis or take more special form.

2. The weight function has a special form.

3. The essential condition $\tau_{n+1} - \tau_n \geq h > 0$ $(n = 1,2,\ldots)$ is imposed on the exponents of $z$.

During 1962-1964, Shen has obtained some theorems generalizing above mentioned results.

Suppose that $L$ is an unbounded curve. It is possible that $L$ is the union of a finite number of connected curves, each of which has a finite number of branches extending to infinity. Furthermore, $L$ has the following properties:

1. It has no loops and is rectifiable in any bounded domain.

2. It divides the entire z-plane into a finite number of unbounded simply-connected regions $G_i$, $i = 1,2,\ldots,n$, each of which contains an angle $\Delta_i$ of magnitude $\frac{\pi}{\alpha_i}$, $\frac{1}{2} \le \alpha_i < +\infty$.

Suppose that on this curve $L$ a real continuous function $p(z)$ is defined such that when $|z|$ is sufficiently large $p(z)$ satisfies

$$p(z) \ge p_0(|z|) = p_0(a) + \int_a^{|z|} \frac{\omega(t)}{t}\,dt \tag{5.1}$$

where $a$ is a constant, $\omega(t) \ge 0$, $\omega(t) \uparrow +\infty$.

Suppose that the sequence $\{\nu_n\}$ of complex numbers satisfies the following conditions:

1. $\lim\limits_{n\to\infty} \frac{n}{|\nu_n|} = D_\nu$;

2. $\operatorname{Re} \nu_n > 0$, $|\operatorname{Im} \nu_n| < C$, $C$ is a constant.

Suppose that $\{\tau_n\}$ is a sequence formed from the distinct elements of $\{\nu_n\}$ and $m_n$, $n = 1,2,\ldots$, the number of repetitions of $\tau_n$ in $\{\nu_n\}$.

THEOREM 1 (Shen [82]). Suppose that in addition to the two properties described above, the curve $L$ also has the third property:

One of the regions $G_i$ $(i = 1,2,\ldots,n)$, say $G_1$, contains a helical angle $P$ with its apex at the origin. At a distance sufficiently large from the origin, $P$ and $\Delta_1$ coincide. Each of the two sides $\ell_1$ and $\ell_2$ of $P$ intersects the circle $|z| = r$, $0 < r < +\infty$, at only one point. At the origin, both $\ell_1$ and $\ell_2$ have tangents. The magnitude of $P$ is greater than or equal to $\frac{\pi}{\alpha_1}$. Furthermore,

$$\frac{1}{\alpha_1} > 2(1-D_\nu). \tag{5.2}$$

Suppose for some $\varepsilon_0 > 0$

$$\int^{\infty} \frac{p_0(r)}{r^{1+\omega}}\,dr = +\infty, \quad \omega = \max\left(\alpha_1,\ldots,\alpha_n, \frac{1}{\frac{1}{\alpha_1} - 2(1-D_\nu)} + \varepsilon_0\right). \tag{5.3}$$

Then the system $\{z^{\tau_n}\log^j z\}$ $(n = 1,2,\ldots,\tau_0 = 0, j = 0,1,\ldots,m_n-1)$ on the $L$ is closed in the class $C[p(z)]$ of functions $f(z)$ which are continuous on $L$ and satisfy the condition

(5.4) $$\lim_{\substack{z\to\infty \\ z\in L}} e^{-p(z)} f(z) = 0$$

i.e.

(5.5) $$\inf_{\{Q(z)\}} \sup_{z\in L} e^{-p(z)} |f(z)-Q(z)| = 0,$$

where the infimum is taken by considering all the linear combinations of the functions $z^{\tau_n}\log^i z$ $(n = 0,1,\ldots,\tau_0 = 0, j = 0,1,\ldots,m_n-1)$.

If the limit $\lim_{n\to\infty} \frac{n}{|\nu_n|}$ does not exist. Under such circumstances, we assume that $\{\nu_n\}$ satisfies the following three conditions:

1. $D_{*\nu} = \underline{\lim}_{n\to\infty} \frac{n}{|\nu_n|} > 0;$

2. $$\sup_{0<\xi<1} \overline{\lim_{r\to\infty}} \frac{N_\nu(v) - N_\nu(r\xi)}{r(1-\xi)} = \tau < +\infty;$$

3. $\operatorname{Re} \nu_n > 0$, $|\operatorname{Im} \nu_n| < C$, $C$ being a constant where $N_\nu(r)$ is the number of $\{\nu_n\}$ lies in $|z| < r$.

Furthermore, by the third condition of $L$ in Theorem 1, $P$ can be considered as an angle formed by two straight lines, with apex at the origin, and magnitude $\frac{\pi}{\alpha_1}$:

(5.6) $$\frac{1}{\alpha_1} > 2(1-D_{*\nu}).$$

With other conditions in Theorem 1 unchanged, we have

THEOREM 2 (Shen [82]). If for $\varepsilon_0 > 0$

(5.7) $$\int^\infty \frac{p_0(r)}{r^{1+\omega}} dr = +\infty, \quad \omega = \max\left(\alpha_1,\ldots,\alpha_n \frac{1}{\frac{1}{\alpha_1} - 2(1-D_{*\nu})} + \varepsilon_0\right)$$

then in the sense of (5.5) the system $\{z^{\tau_n}\log^j z\}$ $(n = 0,1,\ldots,\tau_0 = 0,$ $j = 1,2,\ldots,m_n-1)$ is closed in $C[p(z)]$ on $L$.

The necessity of divergence (5.3) were considered too.

In the paper [82] the approximation is $L_p[p(z)]$ and on a measurable set were also investigated.

If the approximation is considered in a domain, then the closure of polynomial system in an unbounded domain or a domain with disconnected complement, depends not only on the topological properties, but also in an essential way on the measurability of the domain [83], [84].

Suppose that $B_i^*(i = 1,2,\ldots,n)$ are "moon-shaped" bounded domain, i.e. $B_i^*$ are topologically equivalent to the region bounded by two tangent circles, one inside the other. Suppose further that all the $B_i^*$ have the origin as the only common point and the complement of $\overline{B}^* = \sum_{i=1}^{n} \overline{B_i^*}$ is formed by a finite number of domains $G_i$ $(i = 1,2,\ldots,m)$, one of which, say $G_1^*$, contains the point of infinity. Suppose that $G_1^*$ contains a helical angle $P$ with vertex at the origin, whose two sides, each of which has a tangent at the origin, are straight lines at a great distance from the origin. Suppose that each of the two sides $\ell_1$ and $\ell_2$ of $P$ intersects $|z| = r$, $0 < r < +\infty$ at one and and only one point and that the magnitude of $P$ is greater than or equal to $\frac{\pi}{\alpha_1}$. To each $i$ $(i = 1,2,\ldots,m)$ there corresponds a domain of type $\Delta(\alpha_i)$ (it means that it is bounded by the arcs of two circles intersecting at the origin, which form an angle of $\frac{\pi}{\alpha_i}$) which is contained in $G_0^*$.

Suppose that $B^*$ is of class $\{p_0(r)\}$, i.e. when $r$ is sufficiently small, we have

$$\sigma^*(r) \le e^{-p_0(r)}, \tag{5.8}$$

where $\sigma^*(r)$ is the linear measure of $B^*$ on the circle $|z| = r$ and $p_0(r)$ satisfies the condition (5.1).

THEOREM 3 (Shen [85]). Suppose the $\{\nu_n\}$ possesses the above two properties. Under the conditions stated above, if

$$\frac{1}{\alpha_1} > 2(1-D_\nu) \tag{5.9}$$

and

$$\int_0 \frac{p_0(r)}{r^{1-\omega}}\,dr = +\infty, \qquad \omega = \max(\alpha_2,\alpha_3,\ldots,\alpha_m), \tag{5.10}$$

then they system $\{z^{\tau_n}\log^j z\}$ $(n = 1,2,\ldots,j = 0,1,\ldots,m_n-1)$ is closed in $L_z[B^*]$ with respect to $B^*$, i.e., for each $f(z)$

$$\iint_{B^*} |f(z)|^2 dxdg < +\infty \tag{5.11}$$

(we call $f(z) \in L_2[B^*]$) we have

$$\inf_{\{Q\}} \iint_{B^*} |f(z) - Q(z)|^2 dxdy = 0, \tag{5.12}$$

where the infimum is taken with respect to all the polynomials $Q(z)$.

In this paper [85] the closure or the weighted closure of $\{z^{\tau_n}\log^j z\}$ in unbounded domains were also investigated. Furthermore, the necessity of condition in Theorem 3 or in other cases was studied as well.

In paper [86] when the exponents $\{\nu_n\}$ of $z$ is lied in an angle, the closure of system $\{z^{\tau_n}\log^j z\}$ was studied.

By the way we indicate that Dzarbasjan [76] only considered the polynomial approximation.

## §6. THE COMPLETENESS OF THE SYSTEM $\{f(\lambda_n z)\}$

The problem of the completeness of the system of functions $\{f(\lambda_n z)\}$ ($f(z)$ an entire function, and $f^{(n)}(0) \neq 0$, $n = 0,1,2,\ldots$) within the circle $|z| < R$ has been studied in many papers (see for example, Gelfond [87], Markusevic [88], Lohin [89] [90] [91], Boas [92], Leont'ev [93]). It concerns whether or not an arbitrary analytic function $F(z)$ within the circle $|z| < R$ satisfies

$$\inf_{\{Q\}} \sup_{|z|\le R'} |F(z) - Q(z)| = 0, \tag{6.1}$$

where $R'$ is an arbitrary number $0 < R' < R$, and the infimum is taken with respect to all possible finite linear combinations $Q(z)$ of the functions $f(\lambda_n z)$, $n = 1,2,\ldots$ . Dzarbasjan [76] considered the completeness problem of $\{f(\lambda_n z)\}$ on the unbounded domain $G$ in the sense of mean approximation, where $f(z)$ is an entire function; the sequence of Taylor coefficients may include a zero-valued subsequence; the domain $G$ does not include the origin, and is situated outside a certain rectilinear angle with vertex at the origin.

Shen [94] investigated the completeness problem of this system of functions $\{f(\lambda_n z)\}$ on an unbounded domain $G$ in the sense of mean approximation. But unlike all previous papers it was assumed that $f(z)$ is represented by the generalized power series $f(z) = \sum_{n=1}^{\infty} a_n z^{\tau_n}$, where $\{\tau_n\}$ is a sequence of real numbers; and as for the unbounded domain $G$, it was assumed that it is situated outside some curvilinear angle with vertex at the origin.

Let the set $B$ consists of finitely many unbounded simply connected domains $B_i$ $(i = 1,2,\ldots,q)$; then it has everywhere the following two properties:

1) The set $B$ partitions the $z$-plane into finitely many unbounded simply connected domains $G_i$ $(i = 1,2,\ldots,m)$, where every domain $G_i$ respectively includes the rectilinear angle of amplitude $\frac{\pi}{\alpha_i}$, $\frac{1}{2} < \alpha_i < \infty$, and the region $G_1$ contains the curvilinear angle $P$ having some amplitude $\geq \frac{\pi}{\alpha_1}$, the vertex of the angle $P$ being at the origin; its two radical edges being located (when $|z|$ is sufficiently large) on two straight lines; and its two radical edges, at the origin, both having tangents. Furthermore, these two radical edges each intersect any circumference of a circle $|z| = r$, $0 < r < \infty$ in only one point.

2) Let the set $B$ belong to the class $\{p_0(r)\}$, so that the semi-axis $[0,+\infty)$ may be partitioned into two sets $E_1$ and $E_2$ in such a way as to satisfy

1. when $r \in E_1$, $\sigma(r) \leq e^{-p_0(r)}$;

2. when $r$ is sufficient large $(r \geq r_0)$,

$$E(r) \leq e^{-p_0(r)},$$

where $\sigma(r)$ is the linear measure of the intersection of the set $B$ with the circumference of the circle $|z| = r$, and $E(r)$ denotes the linear measure of the intersection of the set $E_2$ with the semi-projecting line $[r,\infty)$, the function $p_0(r)$ having the condition (5.1).

Let $F(s) = f(e^{-s}) = \sum_{n=1}^{\infty} a_n e^{-\tau_n s}$ be a function represented by a Dirichlet series converging on the whole plane, where $a_n \neq 0$ $(n = 1,2,\ldots)$, and $\{\tau_n\}$ be a monotone increasing sequence of positive real numbers; then it has everywhere the density $D$:

$$D = \lim_{n\to\infty} \frac{n}{\tau_n} < +\infty. \tag{6.2}$$

Furthermore, let the (R)-order of $F(s)$ be either less than or equal to $\rho_1$, but assume its type does not exceed $\sigma_1$.

THEOREM (Shen [94]). Under the above mentioned conditions we take $p_0(r) = ur^p$, $u > 0$, $p > 0$ and $0 < \rho_1 \leq p$.

Let the sequence $\{\lambda_n\}$ of complex numbers be located in the angle with vertex at the origin, and with amplitude $2\alpha$, and satisfy one of the following conditions:

When $\rho_1 < p$,

1. $$\lim_{n\to\infty} \frac{n}{\lambda_n^{\frac{p\rho_1}{(p-\rho_1)}}} > \frac{\alpha}{\pi B}\left[(\sigma_1\rho_1)^p\left(\frac{2}{pu}\right)^{\rho_1}\right]^{\frac{1}{\rho-\rho_1}}, \quad \alpha > \frac{b(p-\rho_1)}{p\rho_1},$$

2. $$\lim_{n\to\infty} \frac{n}{\lambda_n^{\frac{p\rho_1}{(p-\rho_1)}}} > \frac{p-\rho_1}{\pi p\rho_1 \cos\frac{\alpha p\rho_1}{p-\rho_1}}\left[(\sigma_1\rho_1)^p\left(\frac{2}{pu}\right)^{\rho_1}\right]^{\frac{1}{\rho-\rho_1}}, \quad \alpha < \frac{\pi(p-\rho_1)}{2p\rho_1},$$

where $B = b \cos b$ is the maximal value of the function $x \cos x$ on $(0,\frac{\pi}{2})$.

When $\rho_1 = p$, then

$$0 < \inf\{|\lambda_n|\} \leq \sup\{|\lambda_n\} \leq \sqrt[p]{\frac{u}{2\sigma_1}} .$$

If

$$\frac{1}{\alpha_1} > 2(1-D) \tag{6.3}$$

and for some number $\varepsilon > 0$

$$\int^\infty \frac{dr}{r^{1+\omega-p}} = \infty, \quad \omega = \max\left(\alpha_1,\alpha_2,\ldots,\alpha_m, \frac{1}{\frac{1}{\alpha_1} - 2(1-D)} + \varepsilon\right), \tag{6.4}$$

then for every class $L_2[B]$ of functions analytic on the interior points of the closed set $\overline{B}$, and satisfying the condition

$$\iint_B |\varphi(z)|^2 d\sigma < \infty, \tag{6.5}$$

we have

$$\inf_{\{Q\}} \iint_B |\varphi(z) - Q(z)|^2 dxdy = 0, \tag{6.6}$$

where the infimum is taken with respect to all possible bounded linear combinations of functions $f(\lambda_n z)$, $n = 1,2,\ldots,$ in other words the system of function $\{f(\lambda_n z)\}$ is complete in the sense of a real mean approximation in the class $L_2[B]$ of functions on $B$.

If (6.3) or (6.4) do not hold, then in general we may say that the theorem is not valid.

If we do not assume that the set $B$ belongs to class $\{e^{-ur^p}\}$ $(u > 0, p > 0)$ (which was the hypothesis in the above theorem), but suppose that it belongs to another class: for example, the class $\{e^{-ur^p/\ln r}\}$ or the class

$\{e^{-ur^p \ell n\, r}\}$, then again we can similarly obtain two theorems [94] for the completeness of $\{f(\lambda_n z)\}$, but some other assumptions were needed in proof.

REFERENCES


1. S. I. Alper, The approximation of functions with a complex variable by polynomials, Izuestia Akad. Nauk USSR Math. ser., 10(1955), 423-444. (Russian).

2. J. G. Chen, Approximation by Cesaro combination of Faber's polynomials on the continuum having fairly smooth boundary, Revue de Math. pures et appliques, 1(1956), 113-146.

3. A. I. Markushevitsch, Theory of analytic functions, M., 1950.

4. C. L. Su, On the best approximation in $E^1$ space of functions by polynomials and rational functions, Journal of Math. Res. and Exp., 3(1982), 41-50 (Chinese).

5. X. C. Shen and Y. R. Lou, On the best approximation by rational functions in the domain of the complex plane. Acta. Mathematica Sinica, 20 : 4(1977), 301-303 (Chinese).

6. F. C. Xing, The best approximation in $H_p^1(p\geq 1)$ space, Proceedings of the 3rd Conference on approximation theory, Oct. 23-30, 1982, Huangshan (1983), 30-37 (Chinese).

7. S. I. Alper, Approximation of analytic functions in mean in the domain, Dokl. Akad. Nauk USSR, 136 : 2(1961), 265-268 (Russian).

8. X. C. Shen, A survey of recent results of the approximation and expansion by rational functions with prescribed poles, Journal of Math. Res. and Exp., 2 : 2(1982), 127-136, 3 : 2(1983), 103-112 (Chinese).

9. X. C. Shen, A survey of recent results of the approximation by rational functions on the interval, Advances in Mathematics, 10 : 1(1981), 24-34, 10 : 2(1981), 81-93 (Chinese).

10. V. N. Rusak, Rational functions as approximation apparatus, Beloruss. Gos. Univ., Minsk, 1979 (Russian).

11. X. C. Shen and Y. R. Lou, On the best approximation by rational functions on the unit circle, Acta. Mathematica Sinica, 20 : 3(1977), 232-235 (Chinese).

12. S. N. Mergelian and M. M. Dzarbasjan, On best approximation by rational functions, Dokl. Akad. Nauk USSR, 99 : 5(1954), 673-675 (Russian).

13. J. E. Andersson and T. Ganelius, The degree of approximation by rational functions with fixed poles, Math. Zeit, 153(1977), 161-166.

14. Q. A. Pekarskii, The degree of rational approximation with preassigned poles, Dokl. Nauk USSR, 21 : 4(1977), 302-304 (Russian).

15. J. Szabados, Structural properties of continuous functions connected with the order of rational approximation II, Acta. Math. Acad. Sci. Hungaricae, 19(1-2) (1968), 95-102.

16. X. C. Shen and Y. R. Lou, On the best approximation by rational functions in the space $H_p(p\geq 1)$, Acta. Scientiarum Naturalium Universitatis Pekinensis 1(1979), 58-72 (Chinese).

17. H. M. Elliott, On approximation to analytic functions by rational functions, Proc. Amer. Math. Soc. 4 : 1(1953), 161-167.

18. J. L. Walsh, Note on degree of approximation to analytic functions by rational functions with preassigned poles, Proc. Nat. Acad. Sci. U.S.A. 42 : 12(1950), 927-930.

19. K. S. Coqialian, On uniform approximation by rational functions in the complex plane, Izvestia Akad. Nauk Armjan SSR ser. math. 11 : 4(1958), 53-77 (Russian).

20. X. C. Shen, On the best approximation by rational functions with preassigned poles, Acta. Mathematica, 21 : 1(1978), 86-90 (Chinese).

21. X. C. Shen, The best approximation by rational functions with preassigned poles in the space $E_p$ $(1 < p \leq +\infty)$, Annals of Mathematics, 1(1980), 51-62 (Chinese).

22. S. N. Mergelian, On some results in the theory of uniform and best approximation by polynomials and rational functions (Russian), in J.L. Walsh's book, "Interpolation and Approximation by rational functions in the complex domain", translated into Russian, Publishing House of foreign culture, Moscow, 1961, 461-496.

23. S. N. Mergelian, Uniform approximation to functions of complex variable, Uspehia Math. Nauk USSR 7 : 2(1952), 31-122 (Russian).

24. G. Fichera, Approximation of analytic functions by rational functions with prescribed poles. Comm. Pure Appl. Math. 23(1970), 359-370.

25. G. Fichera, On the approximation of analytic functions by rational functions, "Topic in analysis", Lecture Notes in Math., 419, 1970, 79-108.

26. G. Fichera, Uniform approximation of continuous functions by rational functions, Ann. Math. Pura Appl. (4) 84(1970), 375-386.

27. S. I. Alper, Approximation by means to analytic functions of class $E_p$, "Investigation of contemporary problems in the theory of functions with a complex variable", Moscow, 1960, 273-286 (Russian).

28. X. C. Shen and Y. R. Lou, On the best approximation by rational functions in the space $E_p$ $(p>1)$, Acta. Scientiarum Naturalium Universitatis Pekinensis, 2(1978), 1-18 (Chinese).

29. X. C. Shen, Approximation and expansion by rational functions in certain class of domain, A Monthly Journal of Sci. 2(1980), 97-102 (Chinese), Kenue Tongbao 3(1980), 97-101 (English).

30. X. C. Shen, On the approximation by rational functions in certain class of domains, Scienta Sinica, 11(1980), 1029-1039 (Chinese), English translation 24 : 8(1981), 1033-1046.

31. A. C. Cordadze, On singular integral with the Cauchy kernel, Trudy Tibles. Math. Institute, 42(1972), 5-17 (Russian).

32. X. C. Shen, A survey of recent results on approximation theory in China, "Multivariat Approximation Theory II", Birkhauser Basel-Boston-Stuttgart, 1982, 385-406.

33. C. K. Chui and X. C. Shen, Degree of rational approximation in digital filter realization in "Rational Approximation and Interpolation", Ed. by Graves-Morris, Saff and Varga, Spring-Verlag (To appear).

34. C. K. Chui and A. K. Chan, Application of approximation theory methods to recursive digiter filter design, IEEE Trans. on ASSP, 30(1982), 18-24.

35. J. Szabados, Rational approximation in complex domain, studie Sci. Math. Hungarian, 4(1969), 335-340.

36. J. Szabados, Rational approximation to analytic functions on an inner part of the domain of analyticity, in "Approximation Theory" ed. by A. Talbot, Acad. Press. New York, 1970, 165-177.

37. V. N. Rusak, Direct methods in rational approximation with free poles, Dokl. Akad. Nauk BSSR, 22(1978), 18-20 (Russian).

38. C. K. Chui and X. C. Shen, Order of approximation by electrostatic fields due to electrons, Constructive Approximation J. (To appear).

39. L. Bers, A nonstandard integral equation with applications to quasi-conformal mappings, Acta. Math. 116(1966), 113-134.

40. M. I. Knopp, A corona theorem for automorphic forms and related results, Amer. J. Math., 91(1969), 599-618.

41. T. A. Metzger, On polynomial density in $A_q(D)$, Proc. Amer. Math.Soc. 44(1974), 326-330.

42. T. A. Metzger and M. Sheigorn, Polynomial approximations in the Bers spaces, Proc. Conf. Univ. Maryland, College Park, Md 1973, 369-377.

43. C. K. Chui, On approximation in the Bers spaces, Proc. Amer. Math. Soc. 40(1973), 438-442.

44. D. J. Neumann, A lower bound for an area integral, Amer. Math. Monthly 79(1972), 1015-1016.

45. J. L. Walsh, Interpolation and approximation by rational functions in the complex domain, Amer. Math. Soc. Colloquium Pub. Vol. XX, 1956.

46. M. M. Dzarbasjan, Theory of Fourier series on rational functions, Izuestia Akad. Nauk Armi SSR Ser. Math - Phy. 9 : 7(1956), 3-28 (Russian).

47. A. A. Ketbalian, Expansion on generalized trigonometric system, Izvestia Akad. Nauk Arm. SSR Ser. Math. 16 : 6(1963), 3-24 (Russian).

48. M. M. Dzarbasjan, On the expansion of analytic functions on the series of rational functions with precribed set of poles, Izuestia Akad. Nauk Arm. SSR Ser. Math-Phy. 10 : 1(1957), 21-29 (Russian).

49. M. M. Dzarbasjan, Expansion on the system of rational functions with preassigned poles, Dokl. Akak. Nauk USSR 143 : 1(1962), 17-20 (Russian).

50. M. M. Dzarbasjan, Expansion on the system of rational functions with preassigned poles, Izvestia Akad. Nauk Arm. SSR Ser. Math. 2 : 1(1967), 3-51 (Russian).

51. K. S. Tymarkin, Expansion of analytic functions on the series of rational functions with preassigned set of poles, Izvestia Akad. Nauk Arm. SSR Ser. Math-Phy., 14 : 1(1961), 9-31 (Russian).

52. A. M. Loukaski, On the Dzarbasjan's system of rational functions for an arbitrary continuum, Siber. Math. J., 15 : 1(1974), 205-211 (Russian).

53. X. C. Shen, On the expansion by means of rational functions in a certain class of domains, Scienta Sinica 3(1981), 257-263 (Chinese), English translation 24 : 11(1981), 1489-1496.

54. K. S. Loukaski, Expansion in a series on the system of rational functions, Math. Sb., 90(132) : 4(1973), 544-557 (Russian).

55. X. C. Shen, An estimation of the remainder of the expansion by rational functions in the space $E_p (1<p\leq+\infty)$, Annals of Mathematics, 2 : 3(1981), 301-310 (Chinese).

56. L. Carleson, An interpolation problem for bounded analytic functions, Amer. Jour. of Math. 80 : 4(1958), 921-930.

57. P. L. Duren, Theory of $H^p$ spaces, Acad. Press New York and London, 1970.

58. H. S. Shapiro and A. L. Shields, On some interpolation problem for analytic functions, Amer. Jour. of Math. 83(1961), 513-532.

59. M. M. Dzarbasjan, Biorthogonal system and solution of interpolation problem with the knots of bounded multiplicity in $H^2$ spaces, Izvestia Akad. Nauk Armsjan SSR ser. Math. 9 : 5(1974), 339-373 (Russian).

60. M. M. Dzarbasjan, Basis of some biorthogonal systems and solution of multiple interpolation problem in $H^p$ spaces over a half-plane, Izvestia Akad. Nauk USSR ser. Math. 42 : 6(1978), 1322-1384 (Russian).

61. M. M. Dzarbasjan, Basis of some biorthogonal systems and solution of multiple interpolation problem in $H_+^p$ spaces, Dokl. Akad. Nauk USSR 234 : 3(1977), 517-520 (Russian).

62. Z. J. Wu, Interpolation in $H_+^p$ spaces, $0 < p < \infty$ and the generalization of Carleson's theorem, M.D. thesis, Beijing, 1984 (Chinese).

63. X. C. Shen, An efficient solution to the problem of multiple interpolation in $H^\infty$ over the upper half-plane, Journal of Approximation Theory and its Applications, 1 : 1(1985) (to appear).

64. V. M. Martirosian, Efficient solution to the problem of multiple interpolation b $H^\infty$ with the help of method of biorthogonality, Izvestia Akad. Nauk Arm. SSR ser. Math. 16 : 5(1981), 339-357 (Russian).

65. X. C. Shen, A moment problem for certain class of system of analytic functions, Acta. Math. Sinica (to appear) (Chinese).

66. H. M. Hairapetian, On the basis of rational functions in the subspace of $H^p$ $(1 < p < \infty)$, Izvestia Akad. Nauk Arm. SSR ser. Math. 8 : 6(1973), 429-449 (Russian).

67. H. M. Hairapetian, On the basis of biorthogonal systems in complex plane, Izvestia Akad. Nauk Arm. SSR ser. Math. 10 : 2(1975), 133-152 (Russian).

68. H. M. Hairapetian, On the basis of rational functions in the subspace of classes $E_p$ $(1 < p < \infty)$, Izvestia Akad. Nauk Arm. SSR ser. Math. 9 : 3(1974), 171-184 (Russian).

69. M. M. Dzarbasjan, Biorthogonal systems of rational functions and representation of Cauchy kernel, Izvestia Akad. Nauk Arm. SSR ser. Math. 8 : 5(1973), 384-408, (Russian).

70. X. C. Shen, On the basis of rational functions in a certain class of domains, Journal of Approximation Theory and its Applications, 1(1984) (to appear).

71. X. C. Shen, On the incompleteness and basis of a system of analytic functions, Scienta Sinica (to appear).

72. X. C. Shen, On the moment problem of a class of analytic function in the complex plane, Proceedings of International Conference on the Approximation Theory and its Applications, St. Johns, Newfoundland, Canada (to appear).

73. A. L. Saginjan, Completeness of families of analytic functions in a complex domain, Soobsc. Inst. of Math. and Meh. Akad. Nauk Arm. SSR 1(1947), 1-59 (Russian).

74. A. L. Saginjan, A method for investigating the completeness of rational functions in domains with disconnected complement and in unbounded domain, Ph.D. Thesis, Moscow State Univ., Moscow, 1944 (Russian).

75. W. H. Fuchs, On the closure of $\{e^{-t} t^{a_\nu}\}$, Proc. Cambridge Philos. Soc. 42(1946), 91-105.

76. M. M. Dzarbasjan, Metric theorems on the completeness and representability of analytic functions, Ph.D. thesis, Moscow State Univ., Moscow, 1948 (Russian).

77. M. M. Dzarbasjan, Some questions of the theory of weighted polynomial approximations in a complex domain, Math. Sb. 36(1955), 353-440 (Russian).

78. S. Mandelbrojt, Series adherentes, regularisation des suites-applications, Gauthier-Villars, Paris, 1952.

79. M. M. Dzarbasjan and I. O. Hacatrjan, On the completeness of the system of functions $\{z^{\lambda_n}\}$ in the complex plane for weighted square approximation, Dokl. Akad. Nauk USSR 110(1956), 914-917 (Russian).

80. A. F. Leont'ev, Completeness of certain systems of analytic functions in infinite regions, Dokl. Akad. Nauk USSR 121(1958), 797-800 (Russian).

81. J. Y. Yu, On the approximation of functions on the positive real axis by generalized polynomials, Acta. Math. Sinica 8(1958), 190-199 (Chinese).

82. X. C. Shen, On the closure of $\{z^{\tau_n}\log^j z\}$ on unbounded curves in the complex plane, Acta. Math. Sinica 13 : 2(1963), 170-192 (Chinese).

83. M. V. Keldys, Sur l'approximation en moyenne quadratique des functions analytiques, Mat. Sb., 5(1939), 391-401.

84. A. L. Saginjan, On approximation in the mean in a complex domain, Izvestia Akad. Nauk USSR ser. Math. 5(1941), 285-296 (Russian).

85. X. C. Shen, On the closure of $\{z^{\tau_n}\log^j z\}$ in a domain of the complex plane, Acta. Math. Sinica 13 : 3(1963), 405-418 (Chinese).

86. X. C. Shen, On approximation of functions in the complex plane by the system of functions $\{z^{\tau_n}\log^j z\}$, Acta. Math. Sinica 14 : 3(1964), 406-414 (Chinese).

87. A. Gel'fond, Sur les systèmes complets de fonctions analytiques, Mat. Sb, 4(1938), 149-156.

88. A. I. Markusevic, Sur les bases l'espace des fonctions analytiques, Mat. Sb. 17(1945), 211-252.

89. I. F. Lohin, On the completeness of a system of analytic functions, Uc. Zap. Gor'kov, Gos. Univ. 28(1955), 24-27 (Russian).

90. I. F. Lohin, On the completeness of a system of functions of the form $\{F(\lambda_n z)\}$, Dokl. Akad. Nauk SSSR, 81(1951), 141-144 (Russian).

91. I. I. Ibragimov, On the completeness of the system of analytic functions $\{F(\alpha_i z)\}$, Izvestia Akad. Nauk SSSR ser. Math. 13(1949), 45-54 (Russian).

92. R. P. Boas, Jr., Fundamental sets of entire functions, Ann. of Math. (2) 47(1946), 21-32.

93. A. F. Leont'ev, On the completeness of a system of analytic functions, Mat. Sb. 31(1952), 381-413 (Russian).

94. X. C. Shen, On the completeness of system of functions $\{f(\lambda_n z)\}$, Acta. Math. Sinica, 14 : 1(1964), 103-118 (Chinese), Scienta Sinica 14 :

Department of Mathematics
Peking University
Beijing

Contemporary Mathematics
Volume 48, 1985

# DIRICHLET SERIES

Yu Chia-yung (Yu Jia-rong)

We state in this paper some aspects of the study of Dirichlet series in China since 1949, not including the study in connection with number theory. Since our information is limited, we do not pretend that this paper is complete in all the related aspects.

## I. ORDINARY DIRICHLET SERIES

1.1. CONVERGENCE. Consider a Dirichlet series

$$f(s) = \sum a_n e^{-\lambda_n s} \quad (0 < \lambda_1 < \lambda_2 < \ldots < \lambda_n \uparrow \infty;\ s = \sigma + it), \tag{1.1}$$

where $a_n$ are complex numbers and $\sigma$ and $t$ are real variables. All kinds of its abscissas of convergence can be calculated or estimated by Kojima-Knopp formula [4],[16] or Valiron formula [33]. Take a sequence $\{\mu_n\}(0<\mu_1<\mu_2<\ldots<\mu_n\uparrow\infty)$ such that $\overline{\lim}(\mu_{n+1}-\mu_n) < \infty$ and $\overline{\lim}(\log n/\mu_n) = D^* < \infty$. In a sense the two formulas mentioned above can be unified in Valiron-Kojima-Knopp formula [45]:

$$\overline{\lim}(\log A_n/\mu_n) \le \sigma_c \le \overline{\lim}(\log A_n/\mu_n) + D^*,$$

where $\sigma_c$ denotes the abscissa of convergence of (1.1),

$$A_n = \sup\{|a_{k_n} + a_{k_n+1} + \ldots + a_m|\} \quad (k_n \le m < k_{n+1}),$$

and $\lambda_{k_n-1} \le \mu_n < \lambda_{k_n}$. There are similar formulas for abscissas of absolute and uniform convergence. These results can be extended to Laplace-Stieltjes transforms, to multiple series and to transforms in higher dimensions [44].

Valiron formula can be extended to Dirichlet series with complex exponets. If we have in (1.1) $\lambda_n = \omega_n e^{i\tau_n}$ $(0 \le \omega_1 \le \omega_2 \le \ldots \le \omega_n \to \infty$, $\tau_n$ being real numbers) and $\overline{\lim}(\log n/\omega_n) < \infty$, then we can evaluate the ranges of convergence and absolute convergence of (1.1) on every straight line passing through $s = 0$ [42]. This is a generalization of one of E. Hille's results [13].

1.2. GROWTH. We consider the cases $\sigma_c = -\infty$ and $\sigma_c = 0$ respectively. Let

$$M(\sigma) = \sup\{|f(\sigma + it)|\} \quad (t \in \mathbb{R}, \sigma > \sigma_c).$$

In the two cases define the order $(R)\rho$ of $f(s)$ respectively as follows:

$$\overline{\lim}[\overset{+}{\log}\ \overset{+}{\log}\ M(\sigma)/(-\sigma)] = \rho \quad (\sigma \to -\infty)$$

and

$$\overline{\lim}[\overset{+}{\log}\ \overset{+}{\log}\ M(\sigma)/\log(1/\sigma)] = \rho \quad (\sigma \to +0)^{\dagger},$$

When $0 < \rho < \infty$, define the type $\tau$ of order $(R)\rho$ as follows:

$$\overline{\lim}[\overset{+}{\log}\ M(\sigma)/e^{-\rho\sigma}] = \tau \quad (\sigma \to -\infty)$$

and

$$\overline{\lim}[\overset{+}{\log}\ M(\sigma)/(1(\sigma)^{\rho}] = \tau \quad (\sigma \to +0).$$

When $\sigma_c = -\infty$, we suppose that

$$\lim(\log n/\lambda_n) = D < \infty. \tag{1.2}$$

In terms of the coefficients and exponents J.F. Ritt (See, for example, [22]) has obtained a necessary and sufficient condition for $f(s)$ to be of order $(R)\rho$. Yu Chia-yung [38] proves that

$$\alpha \le \tau \le (\rho De^{\rho D+1} + 1)\alpha \tag{1.3}$$

if the enitre function $f(s)$ is of type $\tau$ of the order $(R)\rho$ $(0 < \rho < \infty)$, where

$$\alpha = \overline{\lim}(\lambda_n/\rho e)|a_n|^{P/\lambda_n}.$$

Xu Quan-hua [37] improves (1.3), obtains

$$\alpha \le \tau \le e^{PD}\alpha$$

and proves that this inequality cannot be further improved. He obtains similar results for proximate orders (R). Yu Chia-yung [38], [42], [46]; Jin Yi-dan [14]; Wu Min [35]; He Long-zhen [11]; Sun Dao-chun [30] and others studied, respectively, regular and perfectly regular growth, zero, finite and infinite

---

† This was defined independently by K. Nandan [26] and by Yu [49]. But the definition of the former is somewhat complicated in form.

proximate orders (R), (p,q) order (R) and proximate order (R) and growth of entire functions defined by (1.1) with complex coefficients or by multiple series or transforms.

When $\sigma_c = 0$, we suppose that

$$\overline{\lim}(n/\lambda_n) < \infty. \tag{1.4}$$

K. Nandan [26] and Yu Chia-yung [49] find independently that a necessary and sufficient condition for $f(s)$ to be of order $(R)\rho$ in $\sigma > 0$ is $\overline{\lim}(\overset{+}{\log}\ \overset{+}{\log}\ |a_n|/\log \lambda_n) = \rho/(p+1)$, where $\rho/(p+1)$ is replaced by $1$ if $p = \infty$. Yu obtains also some results on finite proximate orders (R) and on regular and perfectly regular growth. Yu Jeou-man [57] improves the above results by replacing the condition (1.4) by

$$\overline{\lim}(\log n/\log \lambda_n) < \infty. \tag{1.5}$$

Yu Chia-yung [50], [52], [53]; Yu Jeou-man [56]; He Long-zhen [12] and Sun Dao-chun [30] studied, respectively, zero and infinite proximate orders (R) and (p,q) order (R) and proximate order (R) and obtained some necessary and sufficient conditions.

If $\{\lambda_n\}$ is not too dense on the real axis, the growth of $f(s)$ in certain horizontal strips or half-strips is the same as in the whole plane or half-plane.

When $\sigma_c = -\infty$, we suppose that

$$\inf_{q>0}\ \overline{\lim_{x\to\infty}}\ [N(x+1)q) - N(xq)]/q = D_1 < \infty \tag{1.6}$$

and

$$\underline{\lim}[\log (\lambda_{n+1} - \lambda_n)]/\lambda_n \geq 0, \tag{1.7}$$

where $N(X)$ = the number of $\lambda_n < x (x > 0)$. Let

$$M(\sigma;t_0,a) = \max_t\{|f(\sigma+it)|\}\ (|t-t_0| \leq a,\ \sigma > \sigma_c).$$

By means of a variant form of Mandelbrojt's inequality we prove that $f(s)$ is of order $(R)\rho$ in every horizontal strip of width $>2\pi D_1$, i.e.

$$\forall t_0 \in \mathbb{R},\ \forall a > \pi D_1,\ \overline{\lim}\{[\overset{+}{\log}\ \overset{+}{\log}\ M(\sigma;\ t_0,\ a)]/(-\sigma)\} = \rho\ (\sigma \to -\infty)$$

if $f(s)$ is of order $(R)\rho$ [40]†. We can deduce some other results on growth

†$D_1$ can be replaced by $\overline{\lim}(n/\lambda_n)$.

of $f(s)$ in horizontal strips.

When $\sigma_c = 0$, we suppose that (1.6) and

$$\underline{\lim}[\log(\lambda_{n+1} - \lambda_n)/\log \lambda_n] > -\infty. \tag{1.8}$$

By means of Anderson and Binmore's inequality we prove that $f(s)$ is of order $(R)\rho$ in every horizontal half-strip of width $>2\pi D_1$ in $\sigma > 0$ if $f(s)$ is of order $(R)\rho$ in $\sigma > 0$ [50], [52].

If $\{\lambda_n\}$ satisfies further conditions, the growth of $f(s)$ on every horizontal line or half-line may be the same in a certain sense. Let

$$M_1(\sigma + it) = \sup_x\{|f(x+it)|\} \quad (x \geq \sigma,\ \sigma > \sigma_c,\ \forall t \in \mathbb{R}).$$

When $\sigma_c = 0$, we suppose that (1.6), (1.8)†† and

$$\sum(1/\lambda_n) < \infty. \tag{1.9}$$

Then $f(s)$ is of order $(R)\rho$ on every half-line in $\sigma > 0$, i.e.,

$$\forall t \in \mathbb{R},\ \overline{\lim}[\overset{+}{\log}\ \overset{+}{\log}\ M_1(\sigma + it)/(-\log \sigma)] = \rho \quad (\sigma \to +0)$$

if $f(s)$ is of order $(R)\rho$ in $\sigma > 0$. There are futher results on growth on horizontal half-lines [54]. For entire functions we have similar results.

1.3 DISTRIBUTION OF VALUES. When $\sigma_c = -\infty$, we suppose that (1.2) holds and that $f(s)$ is of order $(R)\rho > 0$. We can construct a real function $\Omega(x)$ such that the abscissa of convergence $\sigma_c^{\phi}$ of

$$\phi(s) = \sum a_n \Omega(\lambda_n) e^{-\lambda_n s}$$

verifies $0 \leq \sigma_c^{\phi} \leq D$. If $s_0 = \sigma_0 + it_0$ is a singular point along $t = t_0$ from a point in $\sigma > \sigma_c^{\phi}$, then, in the case $0 < \rho < \infty$, there is a horizontal Borel line of $f(s)$ in the strip $|t - t_0| \leq \pi/2\rho$ and, in the case $\rho = \infty$ $t = t_0$ is a Borel line of $f(s)$. Conversely, give a Dirichlet series whose abscissa of convergence is finite and for which (1.2) holds, we can construct an entire function defined by a Dirichlet series for which there is a Borel line corresponding to every singular point of the type mentioned above of the former series [38].

Hence from certain theorems on singularities of Dirichlet series we can deduce some results on Borel lines of entire functions defined by Dirichlet series. For example, if $\overline{\lim}(n/\lambda_n) = \Delta$, $\underline{\lim}(\lambda_{n+1} - \lambda_n) > 0$, then, in the case

---

††(1.6) and (1.8) can be replaced by $\underline{\lim}(\lambda_{n+1} - \lambda_n) > 0$.

$0 < \rho < \infty$, in any horizontal strip of width $(2\Delta+1/\rho)\pi$, there is a horizontal Borel line of $f(s)$ and, in the case $\rho = \infty$, in any horizontal strip of width $2\Delta\pi$ there is a horizontal Borel line. Here we precise in a sense some results of S. Mandelbrojt [22], [23].

The above results can be extended to Laplace-Stieltjes transforms and to Dirichlet series with complex exponents [45], [38].

This research was originated by G. Valiron [33] and continued by Yu Chia-yung and C. Tanaka [31], [32].

When $\sigma_c = 0$, we suppose that (1.6) and (1.8) hold. If $\overline{\lim}\, \sigma \overset{+}{\log} M(\sigma) = \infty(\sigma \to +0)$, then in any interval of length $2\pi D_1$ on $\sigma = 0$, there is a Picard point of $f(s)$. If $f(s)$ is of positive finite order (R) or of certain infinite one, then in any interval mentioned above there is a Borel point of $f(s)$ [50], [52].

1.4. SINGULARITIES. Picard and Borel points are singularities. We give here some other results on singularities of Dirichlet series.

We can extend the theorems of Landau-Carlson and Ostrowski to double Dirichlet series [46] and to Dirichlet series with complex exponents [42]. As at the end of 1.1 let $\lambda_n = \omega_n e^{i\tau_n}$. In particular, if $\{\omega_n\}$ satisifes some conditions similar to (1.6) and (1.7) if $\lim \cos \tau_n$ exists and if $\lim (n/\omega_n) = 0$, then the line of convergence of (1.1) is the natural boundary of $f(s)$ [42].

S. Mandelbrojt's [22], [23] extensions of composition theorems to Dirichlet series are very important. In his extensions, he supposes that the series are bounded or of certain growth except for singularities. He indicates that it would be interesting to reject these hypotheses. Following Mandelbrojt's idea and applying Schttky-Valiron theorem, Yu Chia-yung [39] establishes some composition theorems of Dirichlet series without such hypotheses, but the sets of possible singularities are enlarged. Xie Jian-xin [36] proves a theorem of Hurwitz type and extends some results of S. Agmon, Vl. Bernstein and M. Blambert to theorems of this type.

## II. ASYMPTOTIC DIRICHLET SERIES AND APPLICATIONS

2.1 ADHERENT SERIES AND UNIQUENESS OF ANALYTIC FUNCTIONS. The theory of adherent series is one of the most important contributions of S. Mandelbrojt [21]. An adherent series is an asymptotic Dirichlet series in some curvilinear strip. Under an adherent condition (in the form of a simple integral) the moduli of the coefficients can be estimated by the maximum moduli of the function asymptotically represented by the series on some discs in the strip. By means of these inequalities of Cauchy type various problems in analysis can be studied.

P. Malliavin [20] generalizes a theorem of S. Mandelbrojt and N. Wiener. He consider a function $g(z)$ holomorphic in $x > 0$ and finds for $g(z) \equiv 0$ a necessary and sufficient condition (in the form of a simple integral) on the growth and the distribution on the positive real axis of zeros of $g(z)$. P. Malliavin applies this result to study adherent series and applications and obtains more precise results.

Yu Chia-yung [42] extends adherent series to the case of complex exponents in a convenient form. Following P. Lelong's idea [17] he extends adherent series and Malliavin's theorem to the case of several variables, the adherent condition and Malliavin's conditon remaining <u>in the form of a simple integral</u>.

For example, consider $m$ sequences of positive numbers $\Lambda^{(\alpha)} = \{\Lambda_n^{(\alpha)}\}$, where

$$\lambda_{n+1}^{(\alpha)} - \lambda_n^{(\alpha)} \geq h(\text{const}) > 0 \quad (n = 1,2,\ldots;\ \alpha = 1,\ldots,m).$$

Put $\lambda_n^{(\alpha)}(r) = 2\sum(1/\lambda_n^{(\alpha)})\ (\lambda_n^{(\alpha)} < r)$. Let $\{A_n\}$ be a sequence of positive numbers and let

$$A(\sigma) = \sup_m \{n\sigma - \log A_n\}. \tag{2.1}$$

Then $\exists$ a function $g(z_1,\ldots,z_m)$ $(z_\alpha = x_\alpha + iy_\alpha)$ analytic in $x_1 > 0,\ldots,$ $x_m > 0$, $\not\equiv 0$, and verifying $g(\Lambda^{(1)} \times\ldots\times \Lambda^{(m)}) = \{0\}$ and

$$|g(z_1,\ldots,z_m)| \leq A_n \quad (x_\alpha > 0,\ \sum[x_\alpha] = n;\ \alpha = 1,\ldots,m;\ n = 1,2\ldots)$$

if and only if $\exists\alpha \in \{1,\ldots,m\}$, $b_\alpha \in \mathbb{R}$ such that

$$\int^\infty A(\Lambda^{(\alpha)}(r) - b_\alpha)\frac{dr}{r^2} < \infty.$$

This result can be extended to the case where some $\lambda_n$ are equal [51].

2.2. PROBLEMS OF MOMENTS. The adherent series and Malliavin's theorem can be applied to study uniqueness of solutions of generalized problems of moments in higher dimensions. Applying adherent series we can treat only problems with integral powers. But the results obtained by two methods are different and independent [47], [48], [51]. The Stieltjes-Hamburger problem of moments is introduced by Yu and it contains the Stieltijes and Hamburger problems as special cases.

From the result on the generalized Stieltjes-Hamburger problem of moments we deduce the following theorem:

There exists on

$$E_\ell = \{(t_1,\ldots,t_m)|t_\beta \geq 0,\ t_\gamma \in \mathbb{R}\}\ (\beta = 1,\ldots,\ell;\gamma = \ell+1,\ldots,m;\ell = 0,1,\ldots,m) \tag{2.2}$$

a non-null measure $\mu$ such that

$$\underbrace{\int_0^\infty\cdots\int_0^\infty}_{t}\underbrace{\int_{-\infty}^\infty\cdots\int_{-\infty}^\infty}_{m-\ell} t_1^{n_1}\ldots t_m^{n_m}\,d\mu = 0\ (n_\alpha = 0,1,2\ldots;t_\alpha^0 = 1;\ \alpha = 1,\ldots,m)$$

and

$$\begin{cases}\displaystyle\int_0^\infty\cdots\int_0^\infty\int_{-\infty}^\infty\cdots\int_{-\infty}^\infty t_\beta|d\mu| \leq A_{2n}, \\ \displaystyle\int_0^\infty\cdots\int_0^\infty\int_{-\infty}^\infty\cdots\int_{-\infty}^\infty t_\gamma|d\mu| \leq A_{2n} (n = 0,1,2,\ldots)\end{cases}$$

if any only if the following intergral converges:

$$\int^\infty A_1(\log r)\frac{dr}{r^2}$$

where $A_1(\sigma) = \sup\{2n\sigma - \log A_{2n}\}$.

This theorem contains known results as special cases [10], [29].

2.3. WEIGHTED APPROXIMATION. Applying adherent series with complex exponents $\lambda_n$ we can study weighted approximation by linear combinations of $\{x^{\lambda_n}\}$ on the positive real axis, where $\lambda_n$ are in a half-strip $\{z|x > 0, |y|<k(\text{const.})\}$ and they are not too dense [43]. Under weaker conditions Shen Xie-chang [26], [27] obtains interesting results on weighted approximation by linear combinations of $\{z^{\lambda_n}\log^j z\}$ on curves or on domains in the complex plane. Zhu Fu-liu [58] extends some results of Shen to the case of two complex variables.

Corresponding to the results in 2.2, we can study weighted approximation by generalized polynomials on some regions in a space of higher dimensions. Applying adherent series Bei Sheng-quan [3] studies approximation in the whole plane by lacunary polynomials with integral powers. Applying Malliavin's theorem Yu Chia-yung [47], [51] treats weighted approximation by generalized polynomials with positive powers. A particular case of the results obtained is as follows:

Let $\phi(u)$ be a positive continuous function $(u \geq 0)$ such that $\log\phi(u)$ is a convex function of $\log u$ and that $\forall v > 0,\ u^v/\phi(u) \to 0\ (u \to \infty)$. Then the sequence $\{t_1^{k_1}\cdots t_m^{k_m}\}$ $(k_1,\ldots,k_m = 0,1,2,\ldots)$ is complete in $C_F(E_\ell)$ if and only if

$$\int^\infty \log \phi(u)\frac{du}{u^2} = \infty,$$

where $F(t_1,\ldots,t_m) = \phi(\sqrt{\sum t_\beta + \sum t_\gamma^2})$ $(\beta = 1,\ldots,\ell;\ \gamma = \ell+1,\ldots,m)$, $E_\ell$ is defined in (2.2) and $C_F(E\ )$ is the space of continuous functions $\psi(t_1,\ldots,t_m)$ on $E_\ell$ such that

$$\psi(t_1,\ldots,t_m)/F(t_1,\ldots,t_m) \to 0 \quad (\sum t_\beta + \sum|t_\gamma| \to \infty;\ \ell = 0,1,\ldots,m),$$

and that $\|\psi\| = \max_{E_\ell}|\psi(t_1,\ldots,t_m)/F(t_1,\ldots,t_m)|$.

This is an extension of a result of L.A. Lusternik [19]. In his special case Lusternik has not proved the necessity of the corresponding condition.

2.4. QUASI-ANALYTIC CLASSES. Applying adherent series Bei Sheng-quan gives a sufficient condition for a class of infinitely differentiable functions of two real variables on the whole plane to be a generalized quasi-analytic class. Applying Malliavin's theorem Yu Chia-yung [47] finds some sufficient conditions and necessary ones for a class of infinitely differentiable functions on $\mathbb{R}^m$ or on certain infinite regions to be a generalized analytic class. A particular case is the following theorem:

$\exists$ an infinitely differentiable function $\psi(t_1,\ldots,t_m) \not\equiv 0$ on $E_\ell$ defined in (2.2) such that

$$|\psi^{(n_1,\ldots,n_m)}(t_1,\ldots,t_m)| \le k^n A_n \quad (\sum n_\alpha = n;\ (t_1,\ldots,t_m) \in E_\ell;\ A_n \text{ and } k$$

being positive constants)

and

$$\psi^{(n_1,\ldots,n_m)}(0,\ldots,0) = 0 \quad (n \text{ and } n_\alpha = 0,1,2,\ldots;\ \alpha = 1,\ldots,m;$$

$\ell = 0,1,\ldots,m)$

if and only if the following integral converges:

$$\int^\infty A(\log r)\frac{dr}{r^2},$$

where $A(\sigma)$ is defined by (2.1).

In connection with this problem Ding Shen-rui [5] and Wen Zhi-ying [34] study classification of infinitely differentiable functions of two variables and related properties.

For infinitely differentiable functions on $[0,\infty)$ Deng Guan-tie [7] obtains interesting results. For example, $\exists$ an infinitely differntiable (analytic) function $\psi(x) \not\equiv 0$ on $[0,\infty)$ such that

$$|\psi^{(n)}(x)| \leq k^n A \quad \text{and} \quad \psi^{(\lambda_n)}(0) = 0 (x \in [0,\infty),\ n = 0,1,2,\dots)$$

if and only if $\overline{\lim_{r\to\infty}}\,(2 \sum_{0<\lambda_n<r} (1/\lambda_n) - \log r) < \infty$, where $A$ and $k$ are positive constants and $\lambda_n$ are integers for which $0 = \lambda_0 < \lambda_1 < \lambda_2 < \dots < \lambda_n \uparrow \infty$.

2.5. BOUNDED ANALYTIC FUNCTIONS. The results mentioned in 2.1 can be applied to study uniqueness of bounded analytic functions. Let $F(z)$ be an analytic function in $\Delta = \{z \mid x > -\delta,\ |y| < g(x)\}$ ($\delta$ = const. > 0), where $g(x)$ is a positive continuous function. Yu Chia-yung [41] proves that $F(z) = 0$ if $g(x)$ decreases to zero with $1/x$, if $F(z)$ converges to zero rapidly enough as $z \to \infty$ in $\Delta$ and if $F(z)$ and a part of its derivaties vanish at $z = 0$. Deng Guan-tie [7] generalizes this result by imposing weaker conditons on $g(x)$.

For a bounded analytic function $F(z)$ in $\Delta_1 = \{z \mid x > -\delta,\ |y| < \delta\}$ ($\delta$(const.) > 0) there is a more concise result. Let $\{\lambda_n\}$ be a strictly increasing sequence of integers ($\lambda_0 = 0$). If

$$\overline{\lim_{r\to\infty}}\,[2 \log r - \log_2 r - \log_3 r - \dots - \log_p r - 2 \sum_{0<\lambda_n<r} (1/\lambda_n)] < \infty$$

and

$$\overline{\lim_{t\to\infty}}\,\{[\sum_{r\leq\lambda_n<tr} (1/\lambda_n)]/\log t\} > 0 \quad (\forall r > 0)$$

and if $F^{(\lambda_n)}(0) = 0 (n = 0,1,2,\dots)$, then $F(z) \equiv 0$, where $p$ is an integer $\geq 2$ [55].

Let $F(z)$ be a bounded analytic function in $\Delta_2 = \{f \mid \arg z| < \alpha\} \cup \{0\}$ $(0 < \alpha < \pi)$. If

$$\underline{\lim_{r\to\infty}}\,\{[\sum_{0<\lambda_n<r} (1/\lambda_n)]/\log r\} > 1 - \alpha/\pi$$

and if $F^{(\lambda_n)}(0) = 0 (n = 0,1,2,\dots)$, then $F(z) \equiv 0$ [55]. Deng Guan-tie [6] proves further that $\exists$ a bounded analytic function $F(z) \not\equiv 0$ in $\Delta_2$ such that $F^{(\lambda_n)}(0) = 0 (n = 0,1,2,\dots)$ if and only if $\overline{\lim_{r\to\infty}}\,[\sum_{0<\lambda_n<r} (1/\lambda_n) + (\alpha/\pi - 1)\log r] < \infty$.

The above results can be extended to the case of several variables [55].

## III. DIRICHLET SERIES IN SOME PROBABILITY AND TOPOLOGICAL SPACES

3.1. CONVERGENCE AND GROWTH OF RANDOM DIRICHLET SERIES. Let $\{a_n(\omega)\}$ $(n = 1, 2,\dots)$ be a sequence of complex random variables in the probability space $(\Omega, A, P)$. Let $\{\lambda_n\}$ be a strictly increasing sequence of positive numbers verifying (1.2) with $D = 0$. A Dirichlet series

$$f(s;\omega) = \sum a_n(\omega)e^{-\lambda_n s} \quad (\omega \in \Omega) \tag{3.1}$$

on $(\Omega, A, P)$ is called a random Dirichlet series. Denote the abscissa of convergence and of absolute convergence of (3.1) by $\sigma_c(\omega)$. If $|a_n(\omega)|$ are independent and their distribution functions are $F_n(x)$, then by Borel-Cantelli's lemma,

$$\sigma_c(\omega) = \alpha(-\infty<\alpha<\infty)\text{a.s.} \Leftrightarrow \sum[1-F_n(e^{\lambda n c})]\begin{cases} < \infty & \forall c > \alpha, \\ - \infty & \forall c < \alpha, \end{cases}$$

where $\sigma_c(\omega) = \alpha$ a.s., a.s. means that we have $\sigma_c(\omega) = \alpha$ almost surely, i.e., $P[\sigma_c(\omega) = \alpha] = 1$. For $\sigma_c(\omega) = \pm\infty$ we have similar conditions [49]. In particular, if $F_n(x) = F(x)$ $(n = 1,2,\dots)$ and $F(+0) < 1$, then

$$\sigma_c(\omega) = \begin{cases} 0 \text{ a.s.} \\ \infty \text{ a.s.} \end{cases} \Leftrightarrow \int^{\infty} N(b \log g)\alpha F(g)\begin{cases} < \infty & \forall b > 0; \\ = \infty & \forall b > 0, \end{cases}$$

where $N(x)$ = the number of $\lambda_n < x$ $(x > 0)$. This result contains L. Arnold's zero-one law and its generalizations [1], [2].

Fan Ai-hua [8] applies Valiron-Kojima-Knopp formula in 1.1 and obtains an analogous result in the general case.

Suppose that $\sigma_c(\omega) = 0$ a.s. and that (1.4) or (1.5) holds. Put

$$M(\sigma;\omega) = \sup\{|f(\sigma+it;\omega)|\} \quad (\sigma > 0,\ t \in \mathbb{R}).$$

We have the following theorem:

If $|a_n(\omega)|$ are independent and their districution functions are $F_n(x)$, then

$$\overline{\lim_{\sigma\to+0}}\ [\overset{+}{\log}\ \overset{+}{\log} M(\sigma;\omega)/\log(1/\sigma)] = \rho(0 < \rho < \infty) \text{ a.s.}$$

$$\Leftrightarrow \sum\ [1-F_n(\exp \lambda_n{}^c)]\begin{cases} < \infty & \forall c \in (\rho/(\rho+1),\infty), \\ = \infty & \forall c \in (0,\rho/(\rho+1)). \end{cases}$$

There are analogous results for $\rho = 0$ and $\infty$. In particular, if $F_n(x) = F(x)$ and $F(+0) < 1$, then

$$\overline{\lim_{\sigma\to+0}} [\overset{+}{\log}\overset{+}{\log} M(\sigma;\omega)/\log(1/\sigma)] = \rho (0 < \rho < \infty) \text{ a.s.}$$

$$\Leftrightarrow \int_1^\infty N((\log y)^c)dF(y) \begin{cases} < \infty & \forall c \in (0,1+1/\rho), \\ = \infty & \forall c \in (1+1/\rho,\infty). \end{cases}$$

This result can be regarded as a generalization of L. Arnold's zero-one law since the last integral can be replaced by

$$\int_1^\infty (\log y)^c dF(y)$$

when $\lambda_n = n$.

For zero, infinite and proximate order (R) we have similar conclusions.

When $\sigma_c(\omega) = -\infty$ a.s., Mao Chao-lin [24] obtains similar results.

3.2. DISTRIBUTION OF VALUES. Let $\Omega = \overset{\infty}{\underset{k=1}{X}} \Omega_k$, where $\Omega_k = \{-1,1\}$ or $[0,1]$. Points in $\Omega$ are denoted by $\omega = (\omega_1,\omega_2,\ldots)$.

Define and interval $I$ in $\Omega$ by

$$I = (\overset{n}{\underset{k=1}{X}} I_k) X (\overset{\infty}{\underset{k=k+1}{X}} \Omega_k) \quad \text{for some positive integer } n,$$

where $I_k$ is an open interval in $\Omega_k$ or $\Omega_k$ in the case $\Omega_k = [0,1]$ and it is $\pm 1$ or $\Omega_k$ in the case $\Omega_k = \{-1,1\}$. Starting from $I$ we construct two compact topological space $\Omega$ [53].

On the other hand, let $\Omega_k$ denote a probability space $(\Omega_k, A_k, P_k)$. In the case $\Omega_k = \{-1,1\}$, $A_k$ is composed of all sub-sets of $\{-1,1\}$ and $P_k[\omega_k = 1] = P_k[\omega_k = -1] = 1/2$. In the case $\Omega_k = [0,1]$, $A_k$ is composed of all measurable sets in $[0,1]$ and $P_k[E] = mE$, where $E \in A_k$ and $mE$ denotes Lebesgue measure of $E$. Based on $\Omega_k$ we construct two probability spaces $(\Omega, A, P)$.

In a compact topological space $\Omega$ if a property holds on the intersection of countable dense sets, we say that this property holds quasi-surely (q.s.) on $\Omega$. The almost sure (a.s.) property in a probability space is defined as in 3.1.

Consider the series defined on the probability or topological spaces $\Omega$:

$$f_1(s;\omega) = \sum a_n e^{2\pi\omega_n i} e^{-\lambda_n s} \quad (\omega_n \in [0,1]) \tag{3.1}$$

and

$$f_2(s;\omega) = \sum a_n \omega_n e^{-\lambda_n s} \quad (\omega_n \in \{-1,1\}, \tag{3.2}$$

where $\lambda_n$ and $s$ are the same as in (1.1). Denote the abscissa of convergence of (3.1) and (3.2) by $\sigma_c$.

When $\sigma_c = -\infty$ q. or a.s., we suppose that (1.2) holds. If $f_1(s;\omega)$ and $f_2(s;\omega)$ are entire functions of order (R)$\rho > 0$ q. or a.s., then they have the following properties q.s. or a.s.: In the case $0 < \rho < \infty$, there is a horizontal Borel line of $f_j(s;\omega)$ $(j = 1,2)$ in every horizontal strip of width $\pi/\rho$ and in the case $\rho = \infty$ every horizontal line is a Borel line of $f_j(s;\omega)$ [38], [53].

When $\sigma_c = 0$ q. or a.s., we suppose that (1.4) or (1.5) holds. Then $f_1(s;\omega)$ and $f_2(s;\omega)$ have the following properties q.s. or a.s.: If $\overline{\lim}\, [\overset{+}{\log} |a_n|/\sqrt{\lambda_n}] = \infty$, then every point on $\sigma = 0$ is a Picard point of (3.1) and (3.2). If $f_1(s;\omega)$ and $f_2(s;\omega)$ are of order (R)$\rho > 0$, then every point on $\sigma = 0$ is a Borel point of (3.1) and (3.2) [49], [53].

Mao Chao-lin [24] and Fan Ai-hua [9] obtain some a.s. and q.s. properties of singularities of double Dirichlet series.

## BIBLIOGRAPHY

[1] L. Arnold, Über die Konvergenz einer zufälligen Potenzreihe, J. reine angew. Math., 222(1966), 9-112.

[2] _________, Konvergenzprobleme bei zufälligen Potenzreihe mit Lücken, Math. Z., 92(1966), 356-365.

[3] Bei Sheng-quan, On generalized quasi-analytic classes of functions of two variables and their applications, J. Wuhan Univ. (Nautral Sc. Ed.), 1964, 2, 14-24.

[4] M. Blamber, Sur l'abscisse de convergence simple des séries de Dirichlet générales, Ann. Inst. Fourier, 14(1964), 509-518.

[5] Ding Shan-rui, On equivalence problem of classes of infinitely differentiable functions of two variables, J. Chekiang Univ., 1981, 1, 73-78.

[6] Deng Guan-tie, Uniqueness of some holomorphic functions, submitted to Chin. Ann. of Math.

[7] ________, $\{\nu_n\}$ quasi-analycity of the class $C\{[(\log_p n)^{a_p} \ldots (\log_q n)^{a_p}]^n\}$ of functions on the half-line and application, submitted to J. of Math. (PRC).

[8] Fan Ai-hua, Convergence of random series and limit distribution of zeros of random polynomials, submitted to Chin. Ann. of Math.

[9] ________, Q.s. singularity of double Dirichlet series, submitted to J. Of Math. (PRC).

[10] W. Feller, An Introduction to Probability and its Applications, Vol. 2, John-Wiley & Sons, Inc., New York, 1970.

[11] He Long-zhen, On the (p,q) order (R) and lower (p,q) order (R) of enitre functions defined by Dirichlet series, J. Wuhan Univ. (Natural Sc. Ed.), 1982, 3, 73-89.

[12] ________, On the p(R) order and p(R) type of analytic functions defined by Dirchlet series, J. of Math. (PRC), 3(1983), 61-72.

[13] E. Hille, On Dirichelt series with complex exponents, Ann. of Math., 26(1924), 261-278.

[14] Jin Yi-dan, Proximate order of an entire function of zero order, J. Chekiang Univ., 1979, 4, 93-104.

[15] J.-P. Kahane, Some Random Series of Functions, D.C. Heath & Co., Lexington, Mass., 1968.

[16] K. Knopp, Über die Konvergenz-abszisse des Laplace Integrals, Math. Z., 54(1951), 291-296.

[17] P. Lelong, Entension d'un théorème de Carleman, Ann. Inst. Fourier, 12(1962), 627-641.

[18] Li Zhi-lin, The growth of a function defined by a Laplace-Stieltjes transform in its half-plane of convergence, J. Wuhan Univ. (Nautral Sc. Ed.), 1981, 3, 15-28.

[19] L.A. Lusternik, On polynomial approximation of functions defined on the whole plane, Uspehi Mat. Nauk, 8:1(53)(1953), 161-164.

[20] P. Malliavin, Sur quelques procédés d'interpolation, Acta Math., 93 (1955), 179-255.

[21] S. Mandelbrojt, Séries Adhérentes, Régularisation des Suites, Applications, Gauthier-Villars, Paris, 1952.

[22]________, Séries de Dirichlet, Principes et Méthodes, Gauthier-Villars, Paris, 1969.

[23]________, Selecta, Gautheir-Villars, Paris, 1981.

[24] Mao Chao-lin, On the growth of entire functions represented by random Dirichlet series, J. Math. Res. and Exposition, 4(1984), 63-68.

[25] ________, Properties of singular points of random Dirichlet series of two variables, J. Wuhan Univ. (Natural Sc. Ed.), 1983), 1, 9-14.

[26] K. Nandan, On the lower order of analytic functions represented by Dirichlet series, Rev. Roumain Math. pures appl., 21(1976), 1361-1368.

[27] Shen Zie-chang, Completeness of the system of functions $\{z^{\tau_n}\log^j z\}$ on unbounded burves in the comples plane, Acta Math. Sinica, 13(1963), 170-192.

[28]________, Completeness of the system of functions $\{z^{\tau_n}\log^j z\}$ on certain domains in the complex plane, Acta Math. Sinica, 13(1963), 405-418.

[29] T.A. Shoat and Tamarkin, The Problem of Moments, Math. Survey 1, New York, 1943.

[30] Sun Dao-chun, Growth of Dirichlet series and its coefficients, submitted to Hunan Ann. of Math.

[31] C. Tanaka, Note on Dirichlet series (V). On the integral functions defined by Dirichlet series, Tôhoku Math. J., (2)5(1963), 67-78.

[32]________, Note on Dirichlet serives (XV). ON G. Valiron's method of summation and Borel's directions, Yokohama Math. J., 2(1954), 151-164.

[33] G. Valiron, Entire functions and Borel's directions, Proc. Nat. Acad. Sc. U.S.A., 20(1934), 211-215.

[34] Wen Zhi-ying, Some properties of infinitely differentiable functions of two variables, Chin. Ann. of Math., 4(1983), Ser. A., 57-64.

[35] Wu Min, On entire functions of proximate zero order (R) and of perfectly regular growth, J. of Math. (PRC), 4(1984), 113-120.

[36] Xie Jian-xin, Extension of Hurwitz theorem to Dirichlet series, J. Wuhan Univ. (Natural Sc. Ed.), 1982.

[37] Xu Quan-hua, Types of proximate orders (R) of entire functions represented by Dirichlet series, submitted to Chin. Ann. of Math.

[38] Yu Chia-yung (Yu Jia-rong), Sur les droites de Borel de certaines fonctions entières, Ann. Ec. Norm. Sup., (3) 68(1951), 65-104.

[39] ________, Sur les théorèmes de composition des séries de Dirichlet, Bull. Sc. Math., (2) 75(1951), 69-80.

[40]________, Remarques sur une inégalité de M. Mandelbrojt, Bull. Sc. Math., (2) 75(1951), 107-113.

[41] ________, On some functions holomorphic in an infinte region, Proc. Amer. Math. Soc., 3(1952), 232-236.

[42] ________, On generalized Dirichlet series, Acta Math. Sinica, 5(1955), 295-311.

[43] ________, Approximation of functions on the positive real axis by generalized polynomials, Acta Math. Sinica, 8(1958), 190-199.

[44] ________, On the convergence of the double Dirichlet series and the double Laplace transform, J. Wuhan Univ. (Natural Sc. Ed.), 1962), 1, 1-17.

[45] __________, Borel lines of entire functions defined by Laplace-Stieltjes transforms, Acta Math. Sinica, 13(1963), 471-484.

[46] __________, The asymptotic double Dirichlet series and their applications, Acta Sci. Natur. Schol. Sup. Sinessium (Paris Math. Mech. Astro.), 1 (1965), 233-253; J. Wuhan Univ. (Natural Sc. Ed.), 1963, 2, 1-22.

[47] ________, Uniqueness of some analytic functions of several variables and applications, Acta Math. Sinica, 19(1976), 219-238.

[48] ________, On moment problems in two dimensions, J. Wuhan Univ. (Natural Sc. Ed.), 1977, 1, 71-89.

[49] __________, Some properties of random Dirichlet series, Acta. Math. Sinica, 21(1978), 97-118.

[50] _________, Sur la croissance et la répartiation des caleurs des séries de Dirichlet qui ne convergent que dans un demi-plan, C. R. Acad. Sc. Paris, 288(1979), Sér. A. 891-893.

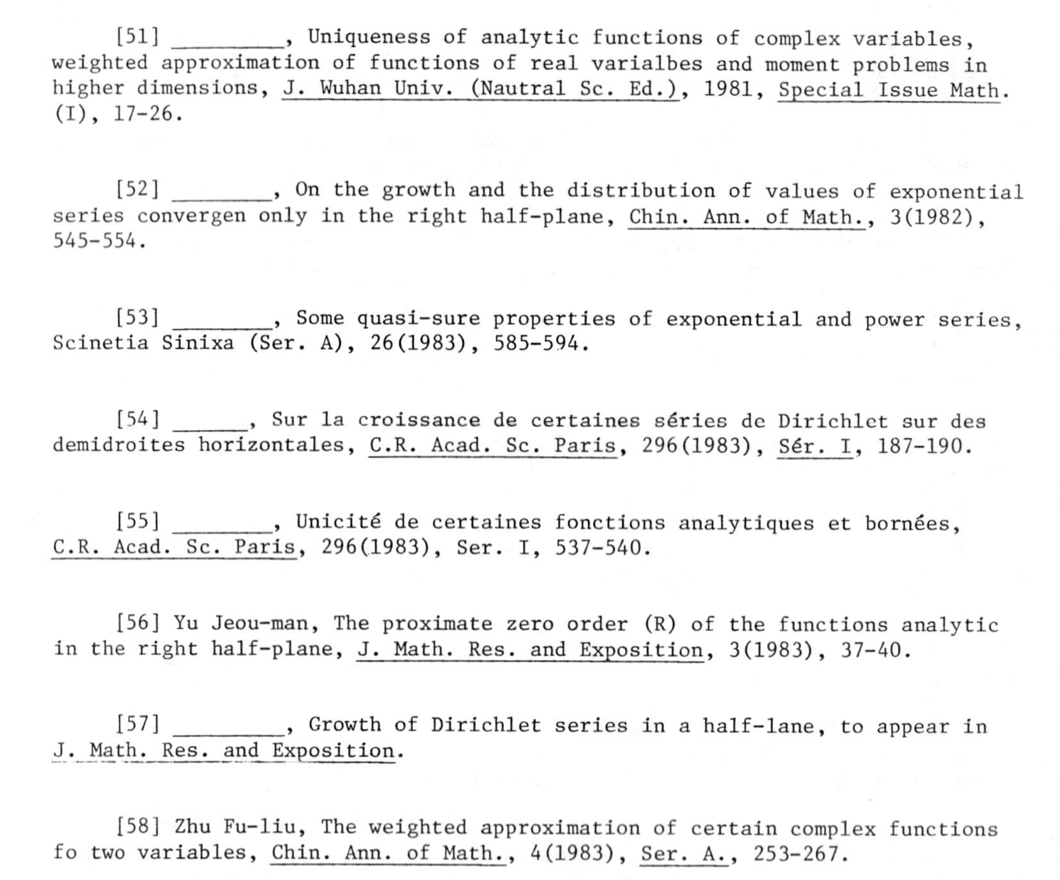

[51] ________, Uniqueness of analytic functions of complex variables, weighted approximation of functions of real varialbes and moment problems in higher dimensions, J. Wuhan Univ. (Nautral Sc. Ed.), 1981, Special Issue Math. (I), 17-26.

[52] ________, On the growth and the distribution of values of exponential series convergen only in the right half-plane, Chin. Ann. of Math., 3(1982), 545-554.

[53] ________, Some quasi-sure properties of exponential and power series, Scinetia Sinixa (Ser. A), 26(1983), 585-594.

[54] ______, Sur la croissance de certaines séries de Dirichlet sur des demidroites horizontales, C.R. Acad. Sc. Paris, 296(1983), Sér. I, 187-190.

[55] ________, Unicité de certaines fonctions analytiques et bornées, C.R. Acad. Sc. Paris, 296(1983), Ser. I, 537-540.

[56] Yu Jeou-man, The proximate zero order (R) of the functions analytic in the right half-plane, J. Math. Res. and Exposition, 3(1983), 37-40.

[57] ________, Growth of Dirichlet series in a half-lane, to appear in J. Math. Res. and Exposition.

[58] Zhu Fu-liu, The weighted approximation of certain complex functions fo two variables, Chin. Ann. of Math., 4(1983), Ser. A., 253-267.

Department of Mathematics
Wuhan University
Wuhan

Contemporary Mathematics
Volume 48, 1985

# APPLICATIONS OF COMPLEX ANALYSIS TO NONLINEAR ELLIPTIC SYSTEMS OF PARTIAL DIFFERENTIAL EQUATIONS

Guo-chun Wen

The purpose of the present article is to give a survey of some works of Chinese mathematicians on nonlinear elliptic systems of partial differential equations in complex form. These works may be classified as follows:

I. Nonlinear elliptic systems of two partial differential equations of first order in complex form.
   A. Homeomorphic solutions.
      1.1. Quasiconformal mapping problem
      1.2. Quasiconformal glue problem
   B. Boundary value problems
      1.3. Compound boundary value problem
      1.4. Discontinuous Riemann-Hilbert problem

II. Nonlinear elliptic equations of second order in complex form
      2.1. Mixed boundary value problems.
      2.2. Poincaré boundary value problem

III. Nonlinear elliptic systems of two partial differential equations of second order in complex form
      3.1. Oblique boundary value problem
      3.2. Riemann-Hilbert boundary value problem

## I. NONLINEAR ELLIPTIC SYSTEMS OF TWO PARTIAL DIFFERENTIAL EQUATIONS OF FIRST ORDER IN COMPLEX FORM.

In this article, we suppose that $D$ is an N+1-connected bounded domain with the boundary $\Gamma \in C^j_\mu (0<\mu<1,\ j=1$ or $2)$. Without loss of generality, we assume that $D$ is a circular domain which contains $z=0$ and is bounded by the N+1 circles $\Gamma_j:|z-z_j|=\gamma_j,\ j=0,1,\ldots,N$, and denote $\Gamma_{N+1}=\Gamma_0$.

We consider a nonlinear system of two partial differential equations of the first order:

$$\Phi_j(x,y,u,v,u_x,u_y,v_x,v_y)=0(j=1,2), \tag{1.1}$$

where the variables $x$, $y$, the unknown functions $u$, $v$ and the functions $\Phi_j(j=1,2)$ are all real. We suppose that the system (1.1) is uniformly elliptic. Set

$$z=x+iy,\ w=u+iv,$$

and define as usual

$$w_z=\frac{1}{2}(w_x-iw_y),\ w_{\bar z}=\frac{1}{2}(w_x+iw_y).$$

Then, under certain conditions, the system (1.1) can be transformed into a single equation in complex form:

$$\begin{cases} w_{\bar z}=F(z,w,w_z),\ F=Q_1w_z+Q_2\overline{w_z}+A_1w+A_2\bar w+A_3, \\ Q_j=Q_j(z,w,w_z),\ j=1,2,\ A_j=A_j(z,w),\ j=1,2,3, \end{cases} \tag{1.2}$$

and the uniform ellipticity condition is expressed by the inequality

$$|F(z,w,U_1)-F(z,w,U_2)|\le q_0|U_1-U_2|, \tag{1.3}$$

where $q_0(0<q_0<1)$ is a constant (cf. Ref. [8]).

We say that equation (1.2) satisfies the <u>condition</u> $C$ in $D$, if the following conditions hold.

1) $Q_j(z,w,U)(j=1,2)$, $A_j(z,w)(j=1,2,3)$ are measurable in $z$ for all continuous functions $w(z)$ and all measurable functions $U(z)\in L_{p_0}(D-\{0\})(z<p_0<p)$ and satisfy

$$L_p[A_j(z,w(z)),\bar D]\le k_0<\infty,\ j=1,2,3, \tag{1.4}$$

where $p_0$, $p$, $k_0$ are positive constants.

2) $Q_j(z,w,U)$, $A_j(z,w)$ are continuous in $w\in E$ (the whole plane) for almost every point $z\in D$ and $U\in E$.

3) Eq. (1.1) satisfies the uniform ellipticity condition (1.3) for almost every point $z \in D$ and $w, U_1, U_2 \in E$.

A. Homeomorphic solutions.

In order to study the homeomorphic solutions of equation (1.2), it is indispensable to add the condition $F(z,w,0) = 0$, so that equation (1.2) becomes

$$w_{\bar{z}} = F(z,w,w_z), \ F = Q_1(z,w,w_z)w_z + Q_2(z,w,w_z)\overline{w_{\bar{z}}}. \tag{1.5}$$

A homeomorphic solution of equation (1.5) is called a quasiconformal homeomorphism.

1.1. Quasiconformal mapping problem

In the particular case $Q_j = Q_j(z)$ or $Q_j = Q_j(z,w)(j=1,2)$, the existence of homeomorphic solutions of equation (1.5), which map a simply connected domain or a multiply connected domain onto a canonical domain were studied by L. Bers and L. Nirenberg [3], B. Boharski [5], S. Pater [18] and Li Zhong [14]. In the case of simply connected domain, M. A. Levrent'ev [12], [13], B. Bojarski and I. Iwanic [6] proved the fundamental theorem of the existence of quasiconformal homeomorphism for the equation (1.5). The existence of such a quasiconformal homeomorphism in the case of multiply connected domain was proved independently by Fang Ai-nong [9], Wen Guo-chun [29], [30], [46] and V. N. Monakhov [17] by using different methods. Here we state only a result in Refs. [30], [41], [46].

THEOREM 1.1. Suppose that the equation (1.5) is uniformly elliptic and satisfies the condition C in an N+1-connected domain D. Then there exists a homeomorphic solution $w(z)$ of equation (1.5) which maps D onto a canonical domain of each of the following types:

1) The N+1-connected circular domain G bounded by N+1 circles $L_j(j=0, 1,\dots,N)$.
2) The rectilinear slit domain G: the boundary of G consists of N+1 rectilinear slits with oblique angles $\theta_j + \frac{\pi}{2}(j=0,\dots,N)$ respectively.
3) The spiral slit domain G: the boundary of G consists of N+1 spiral slits with oblique angles $\theta_j + \frac{\pi}{2}(j=0,\dots,N)$ respectively.
4) The band domain G with N slits.

For proofs of the above theorem, there are many different methods which appeared in the literatures. Applying the Schauder fixed-point theorem, we can prove the above theorem.

1.2. Quasiconformal glue problem

Let $\Gamma_0, \Gamma_1, \ldots, \Gamma_N, L_1, \ldots, L_M$ be the boundary contours of an $N+M+1$ connected domain $D$, where $\Gamma_1, \ldots, \Gamma_N, L_1, \ldots, L_M$ are situated inside $\Gamma_0$. In $D$ there are some mutually exclusive contours $\gamma_1, \ldots, \gamma_n, \ell_1, \ldots, \ell_m$. We assume that

$$\Gamma = \bigcup_{j=0}^{N} \Gamma_j, \quad L = \bigcup_{j=1}^{M} L_j, \quad \gamma = \bigcup_{j=1}^{n} {}_j, \quad \ell = \bigcup_{j=1}^{m} \ell_j \in C_\mu \ (0<\mu<1) \tag{1.6}$$

and denote

$$D_\gamma^- = \bigcup_{j=1}^{n} D_{\gamma_j}, \quad D_\ell^- = \bigcup_{j=1}^{m} D_{\ell_j}, \quad D^- = (D_\gamma^- \cup D_\ell^-) \cap D, \quad D^+ = D \backslash \overline{D^-}. \tag{1.7}$$

where $D_{\gamma_j}$ and $D_{\ell_j}$ are the domains surrounded by $\gamma_j$ and $\ell_j$ respectively.

The general quasiconformal glue problem for the nonlinear complex equation (1.5) in $D^\pm$ is to find a continuous solution $w(z)$ in $\overline{D^\pm} \backslash \{z_0\} (w(z_0) = \infty)$, that satisfies the glue condition

$$\begin{cases} w^+[\alpha(t)] = w^-(t), t \in \gamma, & w^+[\alpha(t)] = \overline{w^-(t)}, \ t \in \ell, \\ w^+[\alpha(t)] = w^+(t), t \in L, & w^+[\alpha(t)] = \overline{w^+(t)}, \ t \in \Gamma, \end{cases} \tag{1.8}$$

where $\alpha(t)$ maps each $\gamma_j$, $\ell_j$, $L_j$ and $\Gamma_j$ topologically onto itself; they give positive shifts on $\gamma \cup \Gamma$ and reverse shifts on $\ell \cup L$, in which $\alpha[\alpha(t)] = t$, for $t \in L \cup \Gamma$, $\alpha(t)$ has the fixed points $a_j \in \Gamma_j$ $j = 0, \ldots, N$, and

$$C'_\mu[\alpha(t), \partial D^\pm] \le d < \infty, \ |\alpha'(t)| \ge d^{-1} > 0. \tag{1.9}$$

The above problem will be called Problem Q.

By using the theorem of conformal glue, the method of elimination, a priori estimates of the homeomorphic solutions for Eq. (1.5) and the Schauder fixed-point theorem, we can obtain the following result.

THEOREM 1.2. Suppose Eq. (1.5) satisfies the condition C, then there exists a homeomorphic solution $w(z) = \begin{cases} w^+(z), z \in D^+ \\ w^-(z), z \in D^- \end{cases}$ of Problem Q for Eq. (1.5), which maps quasiconformally $D^+$, $\overline{D} \cap D_{\gamma_j}$ and $D^- \cap D_{\ell_j}$ onto the domains $G^+$, $G_{\gamma_j}$ $(j = 1, \ldots, n)$ and $G_{\ell_j}$ $(j = 1, \ldots, m)$ respectively, such that $w(z_0) = \infty$.

In addition, we can also prove the quasiconformal glue theorems for each condition of (1.8) for the nonlinear complex equation (1.5) (Cf. Ref. [37]).

B. Boundary value problem

1.3. Compound boundary value problems

The compound boundary value problem for the complex equation (1.2) may be formulated as follows: Find a sectionally regular solution $w^{\pm}(z)$ of (1.2) in $D^{\pm}$, continuous on $\overline{D^{\pm}}$ and satisfying the boundary conditions:

$$w^{+}[\alpha(t)] = G(t)w^{-}(t) + g(t),\ t \in \gamma, \tag{1.10}$$

$$w^{+}[\alpha(t)] = G(t)\overline{w^{-}(t)} + g(t),\ t \in \ell, \tag{1.11}$$

$$w^{+}[\alpha(t)] = G(t)w^{+}(t) + g(t),\ t \in L, \tag{1.12}$$

$$w^{+}[\alpha(t)] = G(t)\overline{w^{+}(t)} + g(t),\ t \in \Gamma, \tag{1.13}$$

where $\alpha(t)$ is defined as in (1.8) and $G(t)$, $g(t)$ satisfy

$$\begin{cases} G(t) \neq 0,\ C_{\nu}[G(t),\partial D^{\pm}] \le d < \infty,\ C_{\nu}[g(t),\partial D^{\pm}] \le d,\ \frac{1}{2} < \nu < 1, \\ G(t)G[\alpha(t)] \equiv 1,\ G(t)g[\alpha(t)] + g(t) \equiv 0,\ t \in L, \\ G(t)\overline{G[\alpha(t)]} \equiv 1,\ G(t)\overline{g[\alpha(t)]} + g(t) \equiv 0,\ t \in \Gamma. \end{cases} \tag{1.14}$$

The above boundary value problem will be denoted by Problem F. Now let

$$K_{\Gamma_j} = \operatorname{Ind} G(t)|_{\Gamma_j} = \frac{1}{2\pi}\Delta_{\Gamma_j} \arg G(t),\ K_{L_j} = \operatorname{Ind} G(t)|_{L_j},\ K_{\gamma_j} = \operatorname{Ind} G(t)|_{\gamma_j},$$
$$K_{\ell_j} = \operatorname{Ind} G(t)|_{\ell_j}$$

and

$$K_{\Gamma} = \sum_{j=0}^{N} K_{\Gamma_j},\quad K_L = \sum_{j=1}^{M} K_{L_j},\quad K_{\gamma} = \sum_{j=1}^{n} K_{\gamma_j},\quad K_{\ell} = \sum_{j=1}^{m} K_{\ell_j}.$$

Besides, denote by $f$ the total of the fixed points of $\alpha(t)$ with $G(t) = -1$ on $L$, and call

$$K = K_{\Gamma} - K_L - f - 2K_{\ell} + 2K_{\gamma} \tag{1.15}$$

the index of Problem F. Without loss of generality, we assume that $K_{\Gamma_j}$ $(j = 1,\dots,N_0)$ are even numbers, and $K_{\Gamma_j}$ $(j = N_0+1,\dots,N)$ are odd numbers.

Since problem F with $K \le 2N-2$ may not have any solution for Eq. (1.2), we discuss the modified boundary value problem for Eq. (1.2), in which the boundary condition (1.13) is represented by

$$w^{+}[\alpha(t)] = G(t)\overline{w^{+}(t)} + g(t) + h(t),\ t \in \Gamma, \tag{1.16}$$

where

$$h(t) = 0, \ t \in \Gamma, \quad \text{when} \quad K \geq 2N-1. \tag{1.17}$$

$$h(t) = \begin{cases} i[G(t)]^{1/2} h_j, \ t \in \Gamma_j, \ 1 \leq j \leq N - K', \ K' = [\frac{K+1}{2}], \\ 0, \ t \in \Gamma_j, \ N - K' < j \leq N + 1, \quad \text{when} \quad 0 \leq K \leq 2N - 2, \end{cases} \tag{1.18}$$

$$h(t) = i[G(t)]^{1/2} h_j, \ t \in \Gamma_j, \ 1 \leq j \leq N + 1, \quad \text{when} \quad K < 0, \tag{1.19}$$

and $h_j (j = 1,\ldots,N+1)$ are all unknown real constants to be determined appropriately, and $h_0 = h_{N+1} = 0$, when $K$ is the negative odd number. Besides, when $K < 0$ we permit that the solution of Eq. (1.2) possesses a pole of order $\leq [\frac{|K|+1}{2}] - 1$ at $z = 0$. Moreover, when $K \geq 0$, we suppose that the solution satisfies the following conditions:

$$\text{Re } G(a_j)^{1/2} \overline{w(a_j)} = b_j, \ j = 1,\ldots,K-N+1, \quad \text{when} \quad K \geq 2N-1, \tag{1.20}$$

$$\text{Re } G(a_j)^{1/2} \overline{w(a_j)} = b_j, \ j = \begin{cases} N_0+1,\ldots,N_0+[\frac{K}{2}]+1, \ \text{when} \ 0 \leq K' \leq N - N_0, \\ N-K'+1,\ldots,N-K'+[\frac{K}{2}]+1, \ \text{when} \ N-N_0<K' \leq N-1, \end{cases} \tag{1.21}$$

where $a_j \in \Gamma_j$, $j = 1,\ldots,N_0$, $a_j (j=N_0+1,\ldots,K-N+1)$ are distinct points on $\Gamma_0 = \Gamma_{N+1}$, $b_j (j=1,\ldots,K-N+1)$ are all real constants, and $|b_j| \leq d$. The above modified boundary value problem for Eq. (1.2) will be called Problem G.

THEOREM 1.3. Let Eq. (1.2) satisfy Condition C. Then Problem G for Eq. (1.2) has a solution $w(z)$ which satisfies the estimates

$$C_\beta[W(z),\overline{D^\pm}] + L_{p_0}[|W_{\bar{z}}| + |W_z|, \overline{D^\pm}] \leq M_1, \tag{1.22}$$

where

$$W(z) = \begin{cases} w(z), \quad \text{when} \quad K \geq -2, \\ w(z)[\zeta(z)]^{K'-1}, \quad \text{when} \quad K < -2, \ K' = [\frac{|K|+1}{2}], \end{cases}$$

and $\zeta(z)$ is a homeomorphism $\beta = 1 - 2/p_0$, $2 < p_0 < \min(p, \frac{1}{1-\nu})$, $M_1 = M_1(q_0,p_0,k_0,D^\pm,\nu,d,\alpha(t))$.

THEOREM 1.4. Under the hypothesis of Theorem 1.3,

1) When $K \geq 2N-1$, Problem F for Eq. (1.2) is solvable.

2) When $0 \leq K \leq 2N-2$, the total of the solvability conditions for Problem F $\leq N - [\frac{K+1}{2}]$.

3) When $K < 0$, Problem F has $N-K-1$ solvability conditions.

In the linear case of Eq. (1.2), the general solution $w(z)$ of problem F $(K \ge 0)$ for Eq. (1.2) admits the representation

$$w(z) = w_0(z) + \sum_{m=1}^{J} c_m w_m(z), \tag{1.23}$$

where $w_0(z)$ is a special solution of Problem F and $w_m(z)(m=1,\ldots,J,$ $J = K-N+1$ for $K \ge 2N-1$, $J = [\frac{K}{2}] + 1 - s(0 \le s \le \min(N - [\frac{K+1}{2}], [\frac{K}{2}]+1)$ for $0 \le K \le 2N-2)$ are the linearly independent solutions of the corresponding homogeneous problem (cf. Ref. [34]).

Furthermore, we also obtain the results of solvability of the boundary value problems for each boundary condition of (1.10)-(1.13) for Eq. (1.2).

For the proof, we introduce a system of four integral equations, and use the continuity method and the Schauder theorem.

1.4. Discontinuous Riemann-Hilbert problem

Now, we study the nonlinear discontinuous Riemann-Hilbert problem with the boundary condition

$$\mathrm{Re}[\overline{\lambda(z)}w(z)] = r[z,w(z)] + s(z),\ z \in \Gamma \tag{1.24}$$

except at the discontinuity points $c_1,\ldots,c_n (N+1 \le n < \infty,\ c_j \in \Gamma_j, j=1,\ldots,N+1)$, where $|\lambda(z)| = 1$, $r(z,w) = r_0(z)r_1(z,w)$, $s(z) = s_0(z)\prod_{j=1}^{n}|z-c_j|^{-\beta_j}$, $0 \le \beta_j/\alpha^2 < 1$, $\alpha = 1 - 2/p_0$, $2 < p_0 < p$. Let $\Gamma^j (j=1,\ldots,n)$ denote all arcs on $\Gamma\backslash\{c,\ldots,c_n\}$ and assume that $\lambda(z), r_0(z), s_0(z), r_1(z,w)$ satisfy

$$\begin{cases} |\hat{r}_1[z(\zeta),0]| \le d,\ C_{\tau_0}\{\hat{r}_1[z(\zeta),\hat{w}[z(\zeta)],L\} \le d + \varepsilon d\{\hat{w}[z(\zeta)],L\}, \\ C_{\tau_0}\{\hat{r}_1[z(\zeta),\hat{w}_1[z(\zeta)]] - \hat{r}_1[z(\zeta),\hat{w}_2[z(\zeta)]],L\} \le \varepsilon d C_{\tau_0}\{\hat{w}_1[z(\zeta)] \\ -\hat{w}_2[z(\zeta)],L\}, L = \zeta(\Gamma), C_\beta[\lambda(z),\Gamma^j] \le d, C_\beta[r_0(z),\Gamma^j] \le d, C_\beta[s_0(z),\Gamma^j] \le d, \end{cases} \tag{1.25}$$

where $\beta(0<\beta<1), \varepsilon(0<\varepsilon<1)d, \tau_0(0<\tau_0<\alpha\min(\beta,\tau))$ are constants, $\hat{w}(z) = \pi(z) \cdot (w(z),\hat{r}_1(z,w(z)) = \pi(z)r_1(z,\hat{w}(z))$ for

$\pi(z) = \prod_{j=1}^{n}|z-c_j|^{\beta_j\alpha^{-2}+\tau}$ $(0<\tau<1), \zeta(z)$ is a homeomorphism of Beltrami equation and $z(\zeta)$ is the reverse function of $\zeta(z)$. We shall find a solution $w(z)$ of Eq. (1.2) which satisfies the boundary condition (1.24) except at points $c_1,\ldots,c_n$ and $w(z) \in W'_{p_0}(D_*)(D_*$ is any compact point set), $\pi(z)w(z) \in C_{\alpha\tau_0}(\overline{D})$ and denote by Problem H this boundary value problem.

Let

$$\frac{\lambda(c_j-0)}{\lambda(c_j+0)} = e^{i\theta_j}, \quad \phi_j = \frac{1}{\pi i}\ln\frac{\lambda(c_j-0)}{\lambda(c_j+0)} = \frac{\theta_j}{\pi} - K_j,$$

$$K_j = \left[\frac{\theta_j}{\pi}\right] + J, \quad J = 0 \text{ or } 1, \; -1 < \phi_j \le 0, \; j = 1,\ldots,n \tag{1.26}$$

and $K = \sum_{j=0}^{n} K_j$ is called the index of Problem H. Applying the method of continuity and the principle of contraction, we can prove the following theorem.

THEOREM 1.5. If Eq. (1.2) satisfies condition C and the constant $\varepsilon$ in (1.25) is sufficiently small, then:

1) When $K \ge 2N-1$, problem H for Eq. (1.2) has a solution $w(z)$.

2) When $0 \le K \le 2N-2$, the total of the solvability conditions for the problem $H \le N - [\frac{K+1}{2}]$ (cf. Ref. [41]).

## II. NONLINEAR ELLIPTIC EQUATION OF SECOND ORDER IN COMPLEX FORM

We know from Ref. [4] that the nonlinear uniformly elliptic equation of second order

$$\Phi(x,y,u_x,u_y,u_{xx},u_{xy},u_{yy}) = 0 \tag{2.1}$$

satisfying some conditions can be transformed into the complex form

$$\begin{cases} u_{z\bar{z}} = F(z,u,u_z,u_{zz},u_{\bar{z}\bar{z}}), \; F = \mathrm{Re}(Qu_{zz}+A_1u_z+A_2u+A_3), \\ Q = Q(z,u,u_z,u_{zz},u_{\bar{z}\bar{z}}), \; A_j = A_j(z,u,u_z,u_{\bar{z}\bar{z}}), \; j = 1,2,3, \end{cases} \tag{2.2}$$

and the uniform ellipticity condition in the domain D can be reduced to the inequality:

$$|F(z,u,u_z,U_1,V_1) - F(z,u,u_z,U_2,V_2)| \le q_0|U_1-U_2|, \tag{2.3}$$

where $q_0(<1)$ is a nonnegative constant.

We suppose that Eq. (2.2) satisfies the condition C, namely

1) $Q(z,u,u_z,U,V)$, $A_j(z,u,u_z,U,V)$ $(j=1,2,3)$ are measurable in $z$ for all continuous functions $u(z)$ and all measurable functions $U(z)$, $V(z)$ in D and satisfy

$$L_p[A_j(z,u(z),u_z,U(z),V(z)),\bar{D}] \le k_0, \; j = 1,2,3 \tag{2.4}$$

where $k_0$ is a nonnegative constant.

2) $Q(z,u,u_z,U,V), A_j(z,u,u_z,U,V)$ are continuous in $u \in R$ (the real axis) and $u_z \in E$ for almost every point $z \in D$ and $U, V \in E$.

3) Eq. (2.2) satisfies the uniform ellipticity condition (2.3) for almost every point $z \in D$ and $u \in R, u_z, U_1, U_2, V_1, V_2 \in E$.

If the coefficient $A_2(z,u(z),u_z,U(z),V(z))$ also satisfies the inequality

$$A_2(z,u(z),u_z,U(z),V(z)) \geq 0 \tag{2.5}$$

for almost every point $z \in D$, the condition C together with (2.5) will be simply called Condition C*.

2.1. Mixed boundary value problems

We first introduce two mixed boundary value problems for Eq. (2.1) or (2.2) satisfying Condition C*, of which one is to find a solution $u(z)$ of Eq. (2.2), which is continuously differentiable on the N+1-connected closed domain $\bar{D}$ and satisfies the boundary condition

$$\frac{\partial u}{\partial \ell} + a_2(t)u(t) = a_3(t) + h(t), \ t \in \Gamma, \quad u(0) = u_0,$$

where $\vec{\ell}$ can be any variable unit vector on the boundary $\Gamma$ provided that $\cos(\ell,n) \geq 0$. $\vec{n}$ being the outward normal vector, $a_j(t)(j=2,3)$ satisfy

$$a_2(t) \geq 0, \ a_2(t) \in C_\mu(\Gamma), \ a_3(t) \in C_\mu(\Gamma), \ 0 < \mu < 1, \ h(t) = \begin{cases} h_0, & t \in \Gamma_0, \\ 0, & t \in \Gamma \backslash \Gamma_0, \end{cases} \tag{2.7}$$

in which $h_0$ is an unknown constant to be determined appropriately. If $\cos(\ell,n) = 0$ and $a_2(t) = 0$ for every point $t \in \Gamma_j$, then we assume that

$$\int_{\Gamma_j} a_3(t)ds = 0, \quad u(a_j) = b_j, \tag{2.8}$$

where $a_j \in \Gamma_j$, $b_j$ is a constant, $1 \leq j \leq N$. The above boundary value problem will be called Problem $M_1$. In particular, if $\cos(\ell,n) = 0$ and $a_2(t) = 0$ for every point $t \in \Gamma$, then Problem $M_1$ is the Dirichlet boundary value problem. If $\cos(\ell,n) > 0$ for every point $t \in \Gamma$, then Problem $M_1$ is the third boundary value problem.

THEOREM 2.1. Suppose that the second order nonlinear equation (2.2) satisfies Condition C*, then Problem $M_1$ of (2.2) has a solution. Furthermore, under more restrictions on Eq. (2.2), uniqueness of the solution of Problem $M_1$ can be proved (cf. Ref. [39]).

Another one of the mixed boundary value problems is to find a solution $u(z)$ of Eq. (2.2), which is continuous in the closed domain $\overline{D}$ and satisfies the boundary condition

$$a_1(t)\frac{\partial u}{\partial \ell} + a_2(t)u(t) = a_3(t) + h(t),\ t \in \Gamma,\ u(0) = u_0, \tag{2.9}$$

where $\vec{\ell}(=\ell_1+i\ell_2)$ can be arbitrarily chosen provided that $\cos(\ell,n) > 0$ on $\Gamma$, $a_1(t)$ and $a_2(t)$ are nonnegative functions satisfying

$$a_1(t) + a_2(t) \geq 1,\ C'_\nu[a_j(t),\Gamma] \leq d,\ C_\nu[\ell_j(t),\Gamma] \leq d,\ |u_0| \leq d, \tag{2.10}$$

in which $\nu(\frac{1}{2}<\nu<1)$, $d(0\leq d<\infty)$ are constants. The above boundary value problem will be called <u>Problem $M_2$</u>. It is clear that the Dirichlet problem and third boundary value problems are special cases of Problem $M_2$.

THEOREM 2.2. Let the nonlinear equation (2.2) satisfy Condition C*. Then Problem $M_2$ of Eq. (2.2) has a solution $u(z)$ and the solution $u(z)$ satisfies the estimates

$$C'_\alpha[u(z),D_*] \leq M_1,\ L_{p_0}[|u_{z\bar{z}}| + |u_{zz}|,D_*] \leq M_2, \tag{2.11}$$

where $D_* = \overline{D} \cap \{\bigcap_{z^*\in\partial\Gamma^*} (|z-z^*|\geq\delta>0)\}$, $\Gamma^* = \{t|a_1(t) = 0\}$ and $2<p_0<\min(p, \frac{1}{1-\nu})$, $\alpha = 1-2/p_0$, $M_j = M_j(q_0,p_0,k_0,\nu,d,D,\Gamma^*,\delta)$, $j = 1,2$. If $\Gamma^*$ is an empty set, we take $D_* = \overline{D}$.

Using the method of continuity, the principle of contraction and Leray-Schauder theorem, we can prove the existence of solutions of Problem $M_2$ for Eq. (2.2) (cf. Ref. [33]).

### 2.2. <u>Poincaré boundary value problem</u>

The discontinuous Poincaré boundary value problem for the nonlinear equation (2.1) or (2.2) <u>(Problem $P_1$)</u> is defined to be the problem of finding a continuously differentiable solution $u(z)$ in $D_* = \overline{D}\backslash\{c_1,\ldots,c_n\}$ that satisfies the boundary condition

$$\mathrm{Re}[\overline{\lambda(z)u_z}] + a_2(z)u(z) = a_3(z),\ z \in \Gamma,\ u(1) = u_0, \tag{2.12}$$

which is the complex form of the boundary condition

$$\frac{\partial u}{\partial \ell}+\sigma(x,y)u(x,y) = \tau(x,y),\ (x,y) \in \Gamma,\ u(1,0) = u_0, \tag{2.13}$$

where $\vec{\ell}$ denotes the vector varying on the boundary $\Gamma$, $\frac{\partial u}{\partial \ell} = \frac{\partial u}{\partial \ell}\cos(\ell,x) + \frac{\partial u}{\partial y}\cdot\cos(\ell,y) = 2\mathrm{Re}[\overline{\lambda(z)u_z}]$, $\overline{\lambda(z)} = \cos(\ell,x) + i\cos(\ell,y)$, $|\lambda(z)| = 1$, $2a_2(z) = \sigma(x,y)$, $2a_3(z) = \tau(x,y)$ and $u_0$, $\lambda(z)$, $a_j(z)(j=2,3)$ satisfy the conditions

$$|u_0| \le d,\ C_\beta[\lambda(z),\Gamma^j] \le d,\ C_\beta[a_2(z),\Gamma^j] \le \varepsilon d,\ C_\beta[a_3(z),\Gamma^j] \le d, \tag{2.14}$$

in which $\Gamma^j$, $\beta$, $d$, $\varepsilon$ are stated as those in §I.

Now let

$$\begin{cases} \dfrac{\lambda(c_j-0)}{\lambda(c_j+0)} = e^{i\theta_j},\ \phi_j = \dfrac{1}{\pi i}\ln\dfrac{\lambda(c_j-0)}{\lambda(c_j+0)} = \dfrac{\theta_j}{\pi} - K_j, \\ K_j = [\dfrac{\theta_j}{\pi}] + J,\ J = 0 \text{ or } 1,\ -1 < \phi_j \le 0,\ j = 1,\dots,n, \end{cases} \tag{2.15}$$

and denote by $K = \sum_{j=0}^{n} K_j$ the index of Problem $P_1$.

THEOREM 2.3. If the second order nonlinear equation (2.2) satisfies condition C and the constant $\varepsilon$ in (2.14) is sufficiently small, then:

1) When $K \ge 2N-1$, Problem $P_1$ for Eq. (2.2) has $N$ solvability conditions.

2) When $0 \le K \le 2N-2$, the total of the solvability conditions for Problem $P_1 \le 2N - [\frac{K+1}{2}]$.

3) When $K < 0$, problem $P_1$ has $2N-K-1$ solvability conditions.

## III. NONLINEAR ELLIPTIC SYSTENS OF TWO PARTIAL DIFFERENTIAL EQUATIONS OF SECOND ORDER IN COMPLEX FORM

In this section, we consider the nonlinear uniformly elliptic complex equation of second order

$$\begin{cases} w_{z\bar{z}} = F(z,w,w_z,\bar{w}_z,w_{zz},\bar{w}_{zz}),F = Q_1w_{zz}+Q_2\bar{w}_{zz}+A_1w_z+A_2w\bar{w}_{\bar{z}}+A_3\bar{w}_z+A_4w_{\bar{z}}+A_5w \\ +A_6\bar{w}+A_7,Q_j = Q_j(z,w,w_z,\bar{w}_z,w_{zz},\bar{w}_{zz}),j=1,2,A_j=A_j(z,w,w_z,\bar{w}_z),j=1,\dots,7, \end{cases} \tag{3.1}$$

which is a complex form of the nonlinear uniformly elliptic system

$$\Phi_j(x,y,u_1,u_2,u_{1x},u_{1y},u_{2x},u_{2y},u_{1xx},u_{1xy},u_{1yy},u_{2xx},u_{2xy},u_{2yy}) = 0,\ j = 1,2 \tag{3.2}$$

(cf. Ref. [42]). By analogy to Eq. (1.2), we assume that Eq. (3.1) in $D$ satisfies Condition C with the main conditions

$$L_p[A_j(z,w(z),w_z,\bar{w}_z),\bar{D}] \le k_j < \infty,\ p > 2,\ j = 1,\dots,7, \tag{3.3}$$

$k_j \le \varepsilon < 1$, $j = 3,\dots,6$, and

$$|F(z,w,w_z,\bar{w}_z,U_1,V_1)-F(z,w,\bar{w}_z,U_2,V_2)| \le q_0|U_1-U_2|+q_0'|V_1-V_2|, \tag{3.4}$$

$q_0 + q_0' < 1$, $q_0' \le \varepsilon$ for almost every point $z \in D$ and $w,w_z,\bar{w}_z,U_j,V_j(j=1,2) \in E$, where $p, k_0, q_0, q_0'$ are nonnegative constants.

3.1. Oblique boundary value problem

We introduce an oblique boundary value problem for second order complex equation (3.1) with the boundary conditions

$$\begin{cases} \operatorname{Re}[\overline{\lambda_1(t) w_t} + \sigma_1(t) w(t)] = \tau_1(t), t \in \Gamma, \\ \operatorname{Re}[\overline{\lambda_2(t) \overline{w}_t} + \sigma_2(t) w(t)] = \tau_2(t), t \in \Gamma, \end{cases} \tag{3.5}$$

where $|\lambda_j(t)| = 1$, $C_\nu[\sigma_j(t), \Gamma] \le \varepsilon d$, $C_\nu[\tau_j(t), \Gamma] \le d$, $j = 1,2$, $\frac{1}{2} < \nu < 1$, $0 < \varepsilon < 1$, and it will be called Problem P.

In order to give the integral expressions of the solutions and to obtain solvability results for the above problem, we propose to study the corresponding modified boundary value Problem $P_*$ for the first order elliptic system

$$U_{1\bar{z}} = F(z, w, U_1, U_2, U_{1z}, U_{2z}),\ U_{2\bar{z}} = \overline{U}_{1z} \tag{3.6}$$

with the following boundary conditions

$$\begin{cases} \operatorname{Re}[\overline{\lambda_1(t)} U_1(t) + \sigma_1(t) w(t)] = \tau_1(t) + h_1(t),\ t \in \Gamma, \\ \operatorname{Re}[\overline{\lambda_2(t)} U_2(t) + \sigma_2(t) w(t)] = \tau_2(t) + h_2(t),\ t \in \Gamma, \end{cases} \tag{3.7}$$

in which

$$h_j(t) = 0,\ t \in \Gamma, \quad \text{for } K_j = \frac{1}{2\pi} \Delta_\Gamma \arg \lambda_j(t) \ge N, \tag{3.8}$$

$$h_j(t) = \begin{cases} h_{jk}, t \in \Gamma_k,\ 1 \le k \le N - K_j, \\ 0\ , t \in \Gamma_k,\ N - K_j < k \le N+1, \end{cases} \quad \text{for } 0 \le K_j < N,\ j = 1 \text{ or } 2, \tag{3.9}$$

$$h_j(t) = \begin{cases} h_{jk}, t \in \Gamma_k,\ k = 1, \dots, N, \\ h_{j0} + \sum_{m=1}^{-K_j - 1} R_e (H_{jm}^+ + iH_{jm}^-) t^m],\ t \in \Gamma_0, \end{cases} \quad \text{for } K_j < 0,\ j = 1 \text{ or } 2, \tag{3.10}$$

where $h_{jk} (k = 0, 1, \dots, N)$, $H_{jm}^\pm (m = 1, \dots, -K_j - 1,\ j = 1,2)$ are all unknown real constants to be determined appropriately. Besides, we require that the solution $[w(z), U_1(z), U_2(z)]$ of Problem $P_*$ satisfies the relation

$$w(z) = w_0 + \int_0^z [U_1(z) + \sum_{j=1}^{n} \frac{d_m}{z - z_j}] dz + \overline{U_2(z)} d\bar{z}, \tag{3.11}$$

where $w_0$ is a complex constant, and the complex constants $d_m (m = 1, \dots, N)$

are appropriately selected such that $w(z)$ in (3.11) is a single-value function in $D$.

THEOREM 3.1. Let $[w(z),U_1(z),U_2(z)]$ be a solution of Problem $P_*$ for the system (3.6) and $w(z) \in C'_\alpha(\bar{D})U_j(z) \in W'_{p_0}(D)$, $\alpha = 1-2/p_0, 2 < p_0 < \min(p, \frac{1}{1-\nu})$. Then $w(z)$ possesses the expression (3.11), in which $U_j(z)(j=1,2)$ can be written in the following form

$$\begin{cases} U_j(z) = \Phi_j(z) + \tilde{T}_j\rho_j, \Phi_j(z) = \frac{1}{2\pi}\int_\Gamma P_j(z,t)r_j(t)d\theta + \phi_j(z), \\ \tilde{T}_j\rho_j = -\frac{1}{\pi}\iint_D [G_{j1}(z,\zeta)\mathrm{Re}\rho(\zeta)-(-1)^j(G_{j2}(z,\zeta)i\mathrm{Im}\,\rho(\zeta)]d\sigma_\zeta, j = 1,2, \end{cases} \tag{3.12}$$

where $r_j(t) = -\mathrm{Re}[\sigma_j(t)w(t)]+\tau_j(t)$, $j=1,2$, $\rho_1(z) = U_{1\bar{z}} = \rho(z)$, $\rho_2(z) = U_{z\bar{z}} = \overline{\rho(z)}$, $P_j(z,t)$ and $G_{jk}(z,\zeta)(j,k=1,2)$ are the Schwarz kernels and Green functions of Problem $P_*$ respectively, $\Phi_j(z)$ and $\phi_j(z)(j=1,2)$ are analytic functions, and $\phi_j(z)$ satisfies the homogeneous boundary condition

$$\mathrm{Re}[\overline{\lambda_j(t)}\,\phi_j(t)] = h_j(t),\ t \in \Gamma,\ j = 1,2. \tag{3.13}$$

In particular, if $D$ is the unit disc $|z| < 1$ and $\lambda_j(t) = t^{K_j}(j=1,2)$, then the solution $w(z)$ can be written in the form

$$\begin{cases} w(z) = w_0 + \int_0^z \Phi_1(z)dz + \overline{\Phi_2(z)}\,d\bar{z} + H\rho, \\ H\rho = \frac{2}{\pi}\iint_D \{\ell n|\zeta-z|\cdot\rho(\zeta) + \frac{1}{2}[g_1(z,\zeta) + \overline{g_2(z,\zeta)}]\overline{\rho(\zeta)}\}d\sigma_\zeta, \end{cases} \tag{3.14}$$

where

$$g_j(z,\zeta) = \begin{cases} \bar{\zeta}^{-2K_j-2}[\ell n(1-\bar{\zeta}z) + \sum_{j=1}^{2K_j+1}\frac{(\bar{\zeta}z)^j}{j}], & \text{for } K_j \ge 0, \\ \zeta^{-2K_j-2}\,\ell_n(1-\bar{\zeta}z), & \text{for } K_j < 0,\ j = 1,2. \end{cases}$$

THEOREM 3.2. Suppose that the nonlinear elliptic system (2.1) satisfies Condition C and the constant $\varepsilon$ in (3.3), (3.4) and (3.5) is sufficiently small, then the corresponding Problem $P_*$ is solvable, and

1) When the indices $K_j \ge N(j=1,2)$, Problem P for (3.1) has $2N$ solvability conditions, and the general solution depends on $2(K_1+K_2-N+2)$ arbitrarily real constants.

2) When $0 \le K_j < N(j=1,2)$, the total of the solvability conditions for Problem P $\le 4N - K_1 - K_2m$ and the solution depends on $K_1+K_2+4$ arbitrarily real constants.

3) When $0 \le K_1 < N$, $K_2 \ge N$ (of $K_1 \ge N$, $0 \le K_2 < N$), the total of the solvability conditions for Problem P $\le 3N - L_1$ (or $3N - K_2$), and the solution depends on $K_1 + 2K_2 - N + 4$ (or $2K_1 + K_2 - N + 4$) arbitrarily real constants.

4) When $K_1 < 0$, $K_2 \ge N$ (or $K_1 \ge N$, $K_2 < 0$), Problem P has $3N - 2K_1 - 1$ (or $3N - 2K_2 - 1$) solvability conditions, and the solution depends on $2K_2 - N + 3$ (or $2K_1 - N + 3$) arbitrarily real constants.

5) When $K_1 < 0$, $0 \le K_2 < N$ (or $0 \le K_1 < N$, $K_2 < 0$), Problem P has $4N - 2K_1 - K_2 - 1$ (or $4N - K_1 - 2K_2 - 1$) solvability conditions, and the solution depends on $K_2 + 3$ (or $K_1 + 3$) arbitrarily real constants.

6) When $K_1 < 0$, $K_2 < 0$, Problem P has $4N - 2K_1 - 2K_2 - 2$ solvability conditions, and the solution depends on two constants.

Using the Leray-Schauder theorem, we can prove the above theorem (cf. Ref. [36]).

3.2. Riemann-Hilbert boundary value problem

The Riemann-Hilbert problem for second order complex equation (3.1) is to find a continuously differentiable solution $w(z)$ in $\overline{D}$, that satisfies the boundary conditions:

$$\begin{cases} \operatorname{Re}[\overline{\lambda_1(t)}U_1(t) + \sigma_1(t)w(t)] = \tau_1(t) + h_1(t), \ U_1 = w_t, \ t \in \Gamma, \\ \operatorname{Re}[\overline{\lambda_2(t)}w(t)] = \tau_2(t) + h_2(t), \ t \in \Gamma, \end{cases} \tag{3.15}$$

where $\lambda_j(t)$, $\sigma_1(t)$, $\tau_j(t)(j=1,2)$ are stated as those in (3.5), and it will be called Problem R.

Similarly, we consider the corresponding modified boundary value Problem $R_*$ for first order elliptic system (3.6) with the following boundary conditions

$$\begin{cases} \operatorname{Re}[\overline{\lambda_1(t)}U_1(t) + \sigma_1(t)w(t)] = \tau_1(t) + h_1(t), \ U_1 = w_t, \ t \in \Gamma, \\ \operatorname{Re}[\overline{\lambda_2(t)}w(t)] = \tau_2(t) + h_2(t), \ t \in \Gamma, \end{cases} \tag{3.16}$$

where $h_j(t)(j=1,2)$ are stated as those in (3.8)-(3.10).

THEOREM 3.3. If $[w(z), U_1(z), U_2(z)]$ is a solution of Problem $R_*$ for (3.6) and $w(z) \in C_\alpha^1(D)$, $U_j(z) \in W_{p_0}^1(D)$, $\alpha = 1 - 2/p_0$, then $w(z)$ possesses the following expression

$$w(z) = \overline{\Phi_2(z)} + \overline{\tilde{T}}_2 U_1 = \overline{\Phi_2(z)} + \overline{\tilde{T}}_2 \Phi_1 + \overline{\tilde{T}}_2 \tilde{T}_1 \rho, \tag{3.17}$$

where $\Phi_j(z)(j=1,2)$ are stated as those in (3.12) and

$$r_2(t) = \tau_2(t),\ r_1(t) = -\mathrm{Re}[\sigma_1(t)w(t)] + \tau_1(t),$$

$$\tilde{T}_j\rho_j = \frac{1}{\pi}\iint_D [G_{j1}(z,\zeta)\mathrm{Re}\rho_j(\zeta) + G_{j2}(z,\zeta)i\mathrm{Im}\rho_j(\zeta)]d\sigma_\zeta,\ j=1,2,$$

$G_{jk}(z,\zeta)(j,k=1,2)$ are Green functions of Problem $R_*$ respectively. In particular, if $D$ is the unit disc $|z| < 1$ and $\lambda_j(t) = t^{K_j}(j=1,2)$, then the solution $w(z)$ can be written in the form

$$\begin{cases} w(z) = V(z) + X(z),\ V(z) = \overline{\Phi_2(z)} + \tilde{\tilde{T}}_2\Phi_1, \\ X(z) = H_1\rho + \overline{Q}_2 H_1\rho, \end{cases} \tag{3.18}$$

where

$$H_j\rho = \frac{2}{\pi}\iint_D [\ell n|1 - \frac{z}{\pi}|\cdot\rho(\zeta) + g_j(z,\zeta)\overline{\rho(\zeta)}d\sigma_\zeta,\ j = 1,2,$$

$g_j(z,\zeta)(j=1,2)$ are the same functions in (3.14), and

$$Q_j\overline{w} = \begin{cases} \dfrac{1}{2\pi i}\displaystyle\int_\Gamma \dfrac{-\overline{w(t)}+tz^{2K_j+1}\,w(t)}{t-z}dt, & \text{for } K_j \geq 0, \\ \dfrac{1}{2\pi i}\displaystyle\int_\Gamma \dfrac{-\overline{w(t)}+\bar{t}^{2K_j-2}\,w(t)}{t-z}dt, & \text{for } K_j < 0, \end{cases} \quad j = 1,2. \tag{3.19}$$

THEOREM 3.4. Suppose that the nonlinear elliptic system (2.1) satisfies Condition C and the constant $\varepsilon$ in (3.3), (3.4) and (3.15) is small enough, then Problem $R_*$ is solvable, and

1) When the indices $K_j \geq N$ $(j=1,2)$, Problem R for (3.1) is solvable, and the general solution depends on $2(K_1+K_2-N+1)$ arbitrarily real constants.

2) When $0 \leq K_j < N$ $(j=1,2)$, the total of the solvability conditions for Problem R $\leq 2N-K_1-K_2$, and the solution depends on $K_1+K_2+2$ arbitrarily real constants.

3) When $K_j < 0$ $(j=1,2)$, Problem R has $2(N-K_1-K_2-1)$ solvability conditions.

In the linear case of the system (3.1), using the Fredholm theorem on the linear operator equation, we can also obtain the results on the solvability of Problem $P_*$ and Problem $R_*$ with weaker conditions.

For the other cases, we can also write down the solvability conditions of Problem R for (3.1).

Besides, we have also studied the solvability of some boundary value problems for nonlinear elliptic systems of several partial differential equations of first order and second order which satisfy certain conditions (cf. Refs. [40], [41]).

## BIBLIOGRAPHY

[1] Begehr, H., Hile, G. N., Riemann boundary value problems for nonlinear elliptic systems, Complex Variables: Theory and Application, 1 (1983), 239-261.

[2] Bers, L., Theory of pseudoanalytic function, New York, 1953.

[3] Bers, L., Nirenberg, L., On a representation theorem for linear elliptic systems with discontinuous coefficients and its applications, Convegno internaz. equazioni lineari alle derivate parziali, Roma, 1955, 111-140.

[4] Bers, L., On linear and nonlinear elliptic boundary value problems in the plane, Convegno internaz. equazioni lineari alle derivate parziali, Roma, 1955, 141-167.

[5] Bojarski, B., Generalized solutions of a system of differential equations of first order and elliptic type with discontinuous coefficients, Mat. Sb., 43 (1957), 451-503.

[6] Bojarski, B., Iwanic, I., Quasiconformal mappings and nonlinear elliptic equations in two variables I, II, Bull. Acad. Polon Sci. Math. Astr. Phys., 22 (1974), 473-484.

[7] Daniljuk, I. I., A problem with an oblique derivative, Sibirsk. Mat. Ž., 3 (1962), 17-55.

[8] Fang, Ainong, Quasiconformal mappings and the theory of functions of systems of nonlinear elliptic partial differential equations of first order, Acta Math. Sinica, 23 (1980), 280-292.

[9] Fang, Ainong, Riemann's existence theorem of nonlinear quasiconformal mappings, Acta Math. Sinica, 23 (1980), 341-353.

[10] Fang, Ainong, On integral operator and (nonlinear mixed) boundary value problem, Scientia Sinica (Series A), 25 (1982), 225-236.

[11] Gilbert, R. P., Buchanan, J. L., First order elliptic systems: A function theoretic approach, 1983.

[12] Lavrent'ev, M. A., On a class of continuous mapping, Math. Sb., 42 (1935), 407-434.

[13] Lavrent'ev, M. A., A fundamental theorem of the theory of quasiconformal mapping of plane regions, Izves. Akad. Nauk SSSR, 12 (1948), 513-554.

[14] Li Zhong, On the existence of homeomorphic solutions of a system of quasilinear partial differential equations of elliptic types, Acta Math. Sinica, 13 (1963), 454-461.

[15] Li Zhong, Wen Guo-chun, On Riemann-Hilbert boundary value problem of linear partial differential equations of the first order, Acta Math. Sinica, 15 (1965), 599-613.

[16] Li Ming-zhong, Generalized Riemann-Hilbert boundary value problem for a system of second order linear elliptic equations, Scientia Sinica, 23 (1980), 280-298.

[17] Monakhov, V. N., Boundary value problems with free boundaries for elliptic systems of equations, 1983.

[18] Parter, S., On mappings of multiply connected domains by solutions of partial differential equations, Comm. pure Appl. Math., 13 (1960), 167-182.

[19] Polozii, G. M., Generalization of the theory of analytic functions of a complex variable. P-analytic and (p,q)-analytic functions and some of their applications, Kiev, 1965.

[20] Schapiro Z., Sur l'existence des représentations quasiconformes, Dokl. Akad. Nauk SSSR, 30 (1941), 690-692.

[21] Tian Mao-ying, The mixed boundary value problem for a class of nonlinear degenerate elliptic equations of second order in the plane, Beijingdaxue Xuebao, 1982, No. 5, 1-13.

[22] Tutschke, W., Partielle komplexe Differentialgleichugen in einer und in mehreren komplexen Variablen, Berlin, 1977.

[23] Tjurikov, E. V., The nonlinear Riemann-Hilbert boundary value problem for quasilinear elliptic systems, Soviet Math. Dokl., 20 (1979), 863-866.

[24] Vekua, I. N., New methods for solving elliptic equations, 1948.

[25] Vekua, I. N., Generalized analytic functions, 1962.

[26] Vinogradov, V. S., On a problem for quasilinear elliptic systems in the plane, Dokl. Akad. Nauk SSSR, 121 (1958), 579-582.

[27] Vinogradov, V. S., A certain method of solution of a boundary value problem for a first order elliptic system on the plane, Dokl. Akad. Nauk. SSSR, 210 (1971), 767-770.

[28] Wen Guo-chun, The properties of solutions and boundary value problems for elliptic systems of first order, Yingyong Shuxue Yu Jisuan Shuxue (Appl. Math. and Num. Math.), 1979, No. 6, 62-69.

[29] Wen Guo-chun, Modified Dirichlet problem and quasiconformal mappings for nonlinear elliptic systems of first order, Kuxue Tongbao (A Monthly J. of Sci.), 25 (1980), 449-453.

[30] Wen Guo-chun, The Riemann-Hilbert problem for nonlinear elliptic systems of first order in the plane, Acta Math. Sinica, 23 (1980), 244-255.

[31] Wen Guo-chun, The singular case of Riemann-Hilbert boundary value problem, Beijingdaxue Xuebao (Acta Sci. Natur. Univ. Peking), 1981, No. 4, 1-14.

[32] Wen Guo-chun, The Poincaré problem with negative index for linear elliptic systems of second order in a multiply connected domain, J. of Math. Res. and Expos., 1981, No. 3, 61-76.

[33] Wen Guo-chun, The mixed boundary value problem for nonlinear elliptic equations of second order in the plane, Proc. of the 1980 Beijing Sym. on Diff. Geom. and Diff. Eq., 1982, 1543-1557.

[34] Wen Guo-chun, On compound boundary value problems with shift for nonlinear elliptic complex equations of first order, Complex Variables: Theory and Application, 1 (1982), 39-59.

[35] Wen Guo-chun, The third boundary value problem for elliptic systems of second order, Chinese Ann. of Math., 4 (1983), 1-12.

[36] Wen Guo-chun, Oblique derivative boundary value problems for nonlinear elliptic systems of second order, Scientia Sinica (Series A), 26 (1983), 113-124.

[37] Wen Guo-chun, Nonlinear quasiconformal glue theorems, Analytic Functions, Springer-Verlag, 458-463.

[38] Wen Guo-chun, Green functions in a multiply connected domain and integral representations of solutions for elliptic boundary value problems, Yingyong Shuxue Yu Jisuan Shuxue, 1982, No. 1, 55-60.

[39] Wen Guo-chen, On representation theorem of solutions and mixed boundary value problems for second order nonlinear elliptic equations with unbounded measurable coefficients, Acta Math. Sinica, 26 (1983), 533-537.

[40] Wen Guo-chen, Nonlinear boundary value problems for elliptic systems of several equations of first order in a multiply connected domain, Beijingdaxue Xuebao, 1983, No. 4, 1-12.

[41] Wen Guo-chen, Nonlinear discontinuous boundary value problems of nonlinear elliptic systems of first order, Beijingdaxue Xuebao, 1985 (to appear).

[42] Wen Guo-chun, Fang Ai-nong, The complex form and some boundary value problems for nonlinear elliptic systems of second order, Chinese Ann. of Math., 2 (1981), 201-216.

[43] Wen Guo-chun, Tai Chung-wei, The oblique derivative boundary value problem for elliptic complex equations of first order in a multiply domain connected domain, Beijingdaxue Xuebao, 1981, No. 3, 19-29.

[44] Wen Guo-chun, Yang Guang-wu, The Poincaré boundary value problem for linear equations of second order, Advan. in Math. 10 (1981), 157-160.

[45] Wen Guo-chun, Liu Qi-feng, The Riemann-Hilbert problem for elliptic systems of fourth order, Scihuan Shiyuan Xuebao (Shuxue Zhuanji), 1981, 90-103.

[46] Wen Guo-chun, Li Shengxun, Xu Keming, On basic theorems of quasiconformal mappings, Hebei Huagong Xueyuan Xuebao (Shuxue Zhuanji), 1980, 20-40.

[47] Wendland, W., Elliptic systems in the plane, London, 1979.

Department of Mathematics
Peking University
Beijing

Contemporary Mathematics
Volume 48, 1985

# RIEMANN ZETA-FUNCTION

Nan-yue Zhang and Shun-yan Zhang

1. The Riemann zeta-function, closely related to the number theory is expressed by the formula

$$\zeta(s) = \sum_{n=1}^{\infty} \frac{1}{n^s}, \tag{1}$$

where $n$ runs through all integers, or by

$$\zeta(s) = \pi_p \left(1 - \frac{1}{p^s}\right), \tag{2}$$

where $p$ runs through all primes. Either of these two formulae may be taken as the definition of $\zeta(s)$, where $s = \sigma+it$ is a complex variable.

The zeta-function first appeared in a work of Euler as a function of a real variable. Euler used it to prove that the sum of the reciprocals of the prime numbers diverges. In a paper read in 1749 Euler asserted that for real $s$

$$\zeta(1 - s) = 2(2\pi)^{-s} \cos \frac{\pi s}{2} \Gamma(s)\zeta(s).$$

This relation was extablished by Riemann in 1859 [1]. It is now called the functional equation of $\zeta(s)$, by means of which the function $\zeta(s)$ can be defined in the half-plane $\sigma \leq 1$ by analytic continuation. Using the function $\zeta(s)$ for complex $S$, Riemann attempted to prove the prime number theorem and, in the course of his investigation, he realized that a study of the complex zeros of the function $\zeta(s)$ might be indispensable. His paper [1], though only ten pages, involves very deep ideas which may be briefly described in the form of six conjectures.

1) $\zeta(s)$ has an infinite of complex zeros in the region $0 \leq \sigma \leq 1$;

2) For $T > 0$, let $N(T)$ denote the number of the zeros of the function $\zeta(s)$ in the region $0 \leq \sigma \leq 1$, $0 < t \leq T$. Then

$$N(T) = \frac{1}{2\pi} T \log T - \frac{1+\log 2\pi}{2\pi} T + O(\log T);$$

3) Let $\rho = \beta+i\gamma$ denote the zeros of $\zeta(s)$ in the strip $0 \leq \sigma \leq 1$.


0271-4132/85 $1.00 + $.25 per page

Then $\sum |\rho|^{-2}$ converges and $\sum |\rho|^{-2}$ diverges.

4) The integral function

$$\zeta(s) = \pi^{-s/2}(s - 1)\zeta(s)\Gamma(\frac{s}{2} + 1)$$

can be expressed as

$$ae^{bs} \prod_{\rho} (1 - \frac{s}{\rho})e^{s/\rho},$$

where $\rho$ runs through all the non-trivial zeros of $\zeta(s)$;

5) All the zeros of $\zeta(s)$ in the strip $0 \leq \sigma \leq 1$ lie on the line $\sigma = \frac{1}{2}$;

6) $\Lambda(n)$ being the Mangoldt function, if we set

$$\Pi(x) = \sum_{2 \leq n \leq} \frac{\Lambda(n)}{\log n}, \quad \Pi_0(x) = \frac{1}{2} (\Pi(x + 0)+\Pi(x - 0)),$$

we have,

$$\Pi_0(x) = \text{li}x - \sum_{\rho} \text{li}x^{\rho} + \int_n^{\infty} \frac{du}{(u^2-1) \log u} - \log 2 \quad (x > 1),$$

where $\text{li}x^{\rho} = \text{li}\, e^{\rho \log x}$ and

$$\text{li}e^{w} = \int_{-\infty+iv}^{u+iv} \frac{e^z dz}{z}, \quad w = u + iv, \; v \lessgtr 0.$$

This is the well known Riemann formula concerning prime numbers.

The conjectures 1), 3), 4) were proved by Hadamard in two papers on entire functions published in 1892 and 1893 [2]. On the other hand, von Mangoldt [3] proved the conjectures 2) and 6). So only the conjecture 5) remains open, which is now called the Riemann hypothesis.

2. Investigation by Chinese mathematicians on the function $\zeta(s)$ began from about 1950. They obtained various results which we state as follows.

Let $N_0(T)$ be the number of the zeros of $\zeta(s)$ of the form $\frac{1}{2}+it\,(0<t\leq T)$. In 1942 A. Selberg [4] proved the inequality

$$N_0(T) > AT \log T.$$

In 1956, S.H. Min, in [5] gave an explicit value of $A = 1/6000$.

In 1974, N. Levinson [6] improved Min's result, and proved

$$N_0(T) > \frac{1}{3} N(T) \tag{3}$$

by a new method involving complicated calculations. In 1979, C.B. Pan [7] simplified the proof of Levinson. In 1980, S.T. Lou and Q. Yao [8] succeeded to replace the constant $\frac{1}{3}$ in (3) by a slightly larger one 0.35.

Concerning the order of $\zeta(\frac{1}{2}+it)$ as $t \to \infty$, Lindelöf made in 1908 the conjecture

$$\zeta(\frac{1}{2} + it) = O(|t|^{\varepsilon}),$$

for every positive $\varepsilon$.

In 1922, Weyl, Hardy and Littlewood proved

$$\zeta(\frac{1}{2} + it) = O(t^{\frac{1}{6}}\log^{\frac{3}{2}}t).$$

Afterwards many estimations of the form

$$\zeta(\frac{1}{2} + it) = O(t^{\alpha}\log^{\beta}t) \tag{4}$$

were found.

For instance, in 1949, S.H. Min [9] proved (4) for $\alpha = 15/92$, and J.K. Chen [10] in 1965 for $\alpha = 6/37$.

In 1981, using the functional equation of $\zeta(s)$, S.Y. Zhang showed that the function $\zeta(s)$ takes each complex value an infinite number of times.

Under the assumption of the Riemann hypothesis, M. Mozep proved in 1974 that for some special points $S_0$ on the line $\sigma = \frac{1}{2}$, the points $(S_0, |\zeta(S_0)|)$ of the surface $u = |\zeta(s)|$ are hyperbolic. N.Y. Zhang and S.Y. Zhang [12] showed that the set of the points on the line $\sigma = \frac{1}{2}$ yielding parabolic points of the surface $u = |\zeta(s)|$ is closed, but for elliptic points or hyperbolic points of the same surface, the corresponding sets on the line $\sigma = \frac{1}{2}$ are open. Furthermore they obtained the formula

$$\sum_{\rho} \frac{1}{\rho(1-\rho)} = 2 + c - \log \pi - \log 2,$$

where $c$ is the Euler constant and $\rho$ runs over all the zeros of $\zeta(s)$ in the region $0 \le \sigma \le 1$.

In another paper [13], they proved the "modula equation"

$$\sum_{n=2}^{\infty} \frac{\Lambda(n)}{\sqrt{n}} e^{-\frac{z}{2}\log^2 n} - \frac{1}{2}\int_0^{\infty} e^{-\frac{z}{8}x^2} \left(\frac{1}{x} - \frac{e^{\frac{3}{4}x}}{e^x-1}\right)dx$$

$$= -\sqrt{\frac{2\pi}{z}} \sum_{\operatorname{Re}\gamma>0} e^{-\frac{\gamma^2}{2z}} + \sqrt{\frac{2\pi}{z}}\, e^{\frac{1}{8z}} - \frac{1}{4}\left(c + \log\frac{8\pi^2}{z}\right)$$

by making use of the following formula of Weil

$$\lim_{T\to\infty}\sum_{|\gamma|<T} M(\rho) = M(0) + M(1) - 2f(0)\frac{\zeta'(\frac{1}{2})}{\zeta(\frac{1}{2})} + \int_0^\infty [f(0) - \frac{f(n)-f(-n)}{2}]\,\frac{e^{\frac{u}{2}}}{\mathrm{sh}u}du$$

$$- \sum_{n=2}^{\infty} \frac{\Lambda(n)}{\sqrt{n}}\{f(-\log n) + f(\log n)\},$$

where $M(s)$ is the Mellin transform of $f(x)$ and $\rho = \beta+i\gamma$ are the non-trivial zeros of $\zeta(s)$, and obtained the expression of $\zeta(s)$:

$$\zeta(s) = \sum_{n=1}^{m} \frac{1}{n^s} + 2^s\left(\frac{1}{(2m+1)^s} + \frac{1}{(2m+2)^s}\right) - \frac{2^s}{\Gamma(s)}\sum_{n=2}^{\infty} \frac{\Lambda(n)}{n^{2m+1}}(\log n)^{s-1}$$

$$- 2^{s-1}\sum_{\rho}\left\{\frac{1}{(2m+1-\rho)^s} + \frac{1}{(2m+1-\bar{\rho})^s}\right\},\ \mathrm{Re}s > 1.$$

In the same paper [13], they introduced the functions

$$\zeta_\rho(s) = \frac{1}{2}\lim_{a\to\frac{1}{2}+0}\sum_{\rho}\left\{\frac{1}{(a-\frac{1}{2}+\rho)^s} + \frac{1}{(a-\frac{1}{2}+\bar{\rho})^s}\right\} = \sum_{\rho}\left(\frac{1}{\rho^s} + \frac{1}{\bar{\rho}^s}\right),$$

$$\zeta_\rho(s) = \lim_{a\to\frac{1}{2}+0}\left\{\sum_{n=2}^{\infty}\frac{\Lambda(n)}{n^{n+\frac{1}{2}}}\log^{-3}n - \Gamma(1-s)(a-\frac{1}{2})^{s-1}\right\},$$

and proved the functional equation:

$$\zeta_\rho(s) - \Gamma(1-s)\{1 - (1-2^{s-1})\zeta(1-s)\} = -2\Gamma(1-s)\zeta_\rho(1-s),\ \mathrm{Re}(s) < 0.$$

Finally we mention some results of N.Y. Zhang on the function $\zeta(s)$. For instance [14], it presents a simple proof of the functional equation of $\zeta(s)$. Another [15] concerning the Stieltjes constants is as follows. It is known that $\zeta(s)$ is analytic except at $s = 1$ which is a simple pole. We have

$$\zeta(s) = \frac{1}{s-1} + \sum_{n=0}^{\infty} A_n(\zeta - 1)^n,$$

where

$$A_n = \frac{(-1)^n}{n!}\gamma_n,$$

and

$$\gamma_n = \lim_{N\to+\infty} \{ \sum_{k=1}^{N} \frac{(\log k)^n}{k} - \frac{(\log N)^{n+1}}{n+1} \}, \ n = 0,1,2,\ldots \tag{5}$$

which are called Stieltjes constants. N.Y. Zhang obtained the formula

$$\gamma_n = \sum_{k=1}^{\infty} (\frac{\log k}{k})^n - (\frac{\log N}{n+1})^{n+1} - (\frac{\log N}{2N})^n + \int_N^{\infty} p_1(x) f_n'(x) dn,$$

where $p_1(x) = x-[x]-\frac{1}{2}$, $f_n(x) = (\log x)^n/x$. Letting $N\to+\infty$, we get (5). On the other hand, setting $N = 1$, he obtained the inequality

$$|\gamma_n| \le \frac{6}{\pi} \cdot \frac{n!}{2^n} .$$

In the paper [16], N.Y. Zhang proved the formula

$$\Gamma(1-\frac{s}{2})\zeta(s) = \sum_{k=1}^{\infty} \frac{(2\pi)^{2k} B_{2k}}{(k-1)!(2k)!(2n-s)} + \frac{\sqrt{\pi}}{s-1} + 2\int_1^{\infty} f(x) x^{1-s} dx,$$

where $s \neq 1,2,4,6,\ldots$, $B_{2k}$ is Bernoulli number and

$$F(x) = \sum_{n=1}^{\infty} \frac{1}{n^2} e^{-x^2/n^2} - \frac{\sqrt{\pi}}{2x} .$$

In another paper [17], he proved the formulas

$$\Gamma(s)\zeta(s) = \frac{2^{\frac{s-3}{2}}}{\sin\frac{1-s}{4}\pi} \int_0^{\infty} \frac{\text{sh}x - \sin x}{\text{ch}x - \cos x} x^{s-1} dx, \ -1 < \text{Re}s < 0$$

$$\Gamma(s)\zeta(s) = \frac{2^{\frac{s-3}{2}}}{\cos\frac{1-s}{4}\pi} \int_0^{\infty} (\frac{\text{sh}x + \sin x}{\text{ch}x - \cos x} - 1) x^{s-1} dx, \ \text{Re}s > 1$$

and gave three different proofs of the functional equation of $\zeta(s)$.

Finally in the paper [18], N.Y. Zhang proved a formula of Ramanujan by a method of Siegel. Then he deduced the formula

$$\zeta(2k+1) = -2\psi_{-k}(\pi) - (2\pi)^{2k+1} \sum_{j=0}^{[\frac{k+1}{2}]} \frac{(-1)^j B_{2j} B_{2k+2-2j}}{(2j)!(2k+2-2j)!}$$

where $\psi_k(\alpha) = \sum_{n=1}^{\infty} \frac{n^{2k-1}}{e^{2n\alpha}-1}$, which gives an expression of the value of $\zeta(s)$ for a positive odd integer. Besides, as consequence of the calculations, some interesting results such as

$$\sum_{n=1}^{\infty} \frac{\mathrm{cth}\pi n}{n^{2k+1}} = -(2\pi)^{2k+1} \sum_{j=0}^{[\frac{k+1}{2}]} \frac{(-1)^j B_{2j} B_{2k+2-2j}}{(2j)!(2k+2-2j)!}$$

$$\sum_{n=1}^{\infty} \frac{e^{2\pi n}}{(e^{2\pi n}-1)^2} = -\frac{1}{8\pi} + \frac{1}{24}$$

were obtained.

BIBLIOGRAPHY

[1] B. Riemann, Über die Anzahl der primzahlen unter einer gegebenen Gröbe, Ges. Math. Werke und Wissenschaftlicher Nachlob, 2 Anfl, 1892, 145-155.

[2] J. Hadamard, Essai sur l'étude des fonction données par leur développement de Taylor, J. Math. pures. Appl (4), 81(1982), 101-186; Etude sur les propriétés des fonctions entiéres et en particulier d'une fonction considerée par Riemann, J. Math. pures. Appl. (4) 9(1893), 173-215.

[3] H. von Mongoldt, Zu Riemann's Abhandlung "Über die Anzahlen der primzahlen unter einer gegebenen Gröbe", J. Reine Angew. Math. 114(1895), 255-305; Zur Verteilung der Nullstellen der Riemannschen Funktion $\zeta(t)$, Math. Ann, 60(1905), 1-19.

[4] A. Selberg, On the zeros of Riemann's zeta-function, Skr. Norske, Vid. Akad. Oslo 10(1942), 1-59.

[5] S.H. Min, On the non-trivial zeros of Riemann's zeta-function, Beijing Daxne Xnebao, 21(1956), 165-189.

[6] N. Levinson, More than one third of zeros of Riemann's zeta-function are on $\sigma = \frac{1}{2}$, Advances in Math. 131 (1974), 383-436.

[7] C.B. Pan, A Simplications of the proof of Levinson's Theorem, Acta Math. Sinica, Vol. 22(1979), No. 3., 344-353.

[8] S.T. Lou and Q. Yao, Lower bound for zeros of Riemann's zeta function on $\sigma = \frac{1}{2}$, Kexue Tongbao, 25(1980), NO. 7, 292-295.

[9] S.H. Min, On the order of $\zeta(\frac{1}{2}+it)$, Trans. Amer. Math. Soc. 65(1949), 448-472.

[10] J.R. Chen, On the order of $\zeta(\frac{1}{2}+it)$, Acta. Math. Sinica. Vol. 15(1965), No. 2, 159-173.

[11] S.Y. Zhang, An Application of the Functional Equation of $\zeta(s)$, Beijing Daxue Xuebao, 1981, No. 2, 42-46.

[12] N.Y. Zhang and S.Y. Zhang, Two consequences of Riemann Hypothesis, Beijing Daxue Xuebao, 1982, No. 4, 1-6.

[13] N.Y. Zhang and S.Y. Zhang, Weil's formula and Riemann zeta function, Beijing Daxue Xuebao, 1984, NO. 2, 13-19.

[14] N.Y. Zhang, On the Functional Equation of Zeta Function, Beijing Daxue Xuebao, 1982, No. 2, 30-34.

[15] N.Y. Zhang, On the Stieltjes constants of the zeta function, Beijing Daxue Xuebao, 1981, No. 4, 20-24.

[16] N.Y. Zhang, A representation of Riemann zeta-function, Journal of Mathematical Research and Exposition, 1982, No. 4, 119-120.

[17] N.Y. Zhang, The series $\sum_{n=1}^{\infty} \frac{1}{n^2} e^{-z^2/n^2}$ and Riemann zeta function, Acta Math. Sinica, 26(1983), No. 6, 736-744.

[18] N.Y. Zhang, Ramanujan's formula and the Values of Riemann Zeta Function at all positive odd Integers, Advances in Math., 121(1983), No. 1, 61-71.

DEPARTMENT OF MATHEMATICS
PEKING UNIVERSITY
Beijing

Contemporary Mathematics
Volume 48, 1985

# APPLICATION OF THE THEORY OF FUNCTIONS OF A COMPLEX VARIABLE TO CELESTIAL MECHANICS

Zhao-hua Yi

The most important topic of celestial mechanics is to study the motions of celestial bodies (including planets, satellites, stars and stellar systems, etc.). The typical equations of motion of celestial mechanics are taken by the Hamiltonian canonical system as follows:

$$\frac{dx}{dt} = \frac{\partial F}{\partial y}, \quad \frac{dy}{dt} = -\frac{\partial F}{\partial x}, \tag{1}$$

where $x$, $y$ are $n$-dimensional vectors, $F = F(x,y,t)$ is the Hamiltonian function expressed in the form

$$F = F_0 + \mu F_1 + \mu^2 F_2 + \dots . \tag{2}$$

The $\mu$ is called the small parameter. When $\mu = 0$, and $F = F_0$, system (1) is integrable. For example, the well-known three-body problem has $n = 18$, and can be reduced to $n = 6$ by means of the first integrals. $F_1, F_2, \dots$ are called disturbing functions.

There are three different approaches (analytical, qualitative and numerical) to obtaining the results of motion.

1. Analytical approach is to get the asymptotic solution of system (1) with given accuracy and with time $t$ as long as possible. Chinese experts got some interesting results in this field.

(1) Improvement of the convergence in the development of disturbing functions. One of the important processes for obtaining the asymptotic solution is to develop the disturbing functions $F_i$ into the power series for some small quantities, and into trigonometric series for some angular variables (depending on time $t$). In general, the eccentricities $e$, $e'$ and $\sin \frac{I}{2}$ may be selected as those quantities, where $I$ is the relative inclination of osculating orbits of celestial bodies.

According to the idea of Soubbotin [1], Yi, et al. obtained some more general results as follows [2,3,4].

When $e, e'$ and $I$ are equal to zero, the kernel problem reduces to develop the function

$$\Delta_0^k = (1+\alpha^2-2\alpha\cos H)^{-k/2} \tag{3}$$

(where $0 < \alpha < 1$, $k$ is an odd positive integer) into the Fourier series of $H$, and power series of $\alpha$.

The classical results are taken as

$$\Delta_0^k = \sum_{n=-\infty}^{+\infty} b_n^{(k)}(\alpha)\cos nH. \tag{4}$$

where $b_n^{(k)}(\alpha)$ are so-called Laplace coefficients which can be developed into the power series of $\alpha$ with first term $\alpha^n$. The expression (4) will be convergent slowly when $\alpha$ is near to unity. One can use some mapping between complex variables to improve the convergence.

We write (3) as

$$\Delta_0^k = [(1-\alpha z)(1-\alpha z^{-1})]^{-k/2}, \tag{5}$$

in which $z = \exp(iH)$, $i = \sqrt{-1}$. Then (4) becomes the Laurent Series:

$$\Delta_0^k = \sum_{n=-\infty}^{+\infty} b_n^{(k)}(\alpha)z^n,$$

which is convergent in the region $\alpha < |z| < \alpha^{-1}$, because $0, \alpha, \alpha^{-1}, \infty$ are singular points of expression (5). When $\alpha \to 1$, $z = \exp(iH)$ will be near the singular point. This is the reason due to bad convergence.

We selected some conformal mappings such as

$$w = \frac{z-\alpha}{1-\alpha z}, \qquad w = \exp(iS). \tag{7}$$

$$w = \frac{z(z-\alpha)}{1-\alpha z}, \qquad w = \exp(iS). \tag{8}$$

from $z$ to $w$, or $H$ to $S$.

In the case of (7), the Laurent series (6) becomes

$$\Delta_0^k = \sum_{n=-\infty}^{+\infty} B_n^{(k)}w^n = \sum_{n=-\infty}^{+\infty} B_n^{(k)} \cos nS. \tag{9}$$

It is convergent in the region $\beta < |z| < \beta^{-1}$, where

$$0 < \beta = \frac{\alpha}{1+\sqrt{1-\alpha^2}} < \alpha.$$

Therefore, the development (9) along $|w| = 1$ is farther from the singular points $(\beta, \beta^{-1})$ than (6). It means that the convergence will be improved.

In the case of (8), the development corresponding to it is the power series of $\beta^2$.

Then we extended the results of mapping (7) to the cases of $e \neq 0$, $e' \neq 0$ and $I \neq 0$.

(2) Analytical solution of the Hill equation. Dong [5] studied the solution of the Hill equation

$$\frac{d^2x}{dt^2} + \sum_{n=0}^{\infty} (a_n \cos 2nt)x = 0, \tag{10}$$

which always appears in celestial mechanics (for instance, in the Lunar theory), by means of the tree graph method. He obtained the explicit expression of the solution - Hill functions, represented by Taylor-Fourier series of new type. His method might be used to study the solution of non-Fuchsian equations.

2. Numerical approach. One can get the accurate solutions in the interval $(0,t)$, by means of the numerical method in the computer, when $t$ is not too large. But it will run into trouble while the orbit is passing near the singular points, in general, the collision cases.

The singular points of binary collision can be removed by some conformal mappings - called regularized mappings.

For example, the planar restricted problem of three bodies can be written as the complex coordinates $z = x + iy$, where $x$ and $y$ are the cartesian coordinates of the body with infinitesimal mass. Its equation of motion has the form

$$\frac{d^2z}{dt^2} + 2i\frac{dz}{st} = \text{grad}_j \Omega = \frac{\partial\Omega}{\partial x} + i\frac{\partial\Omega}{\partial y} \tag{11}$$

where

$$\Omega = \frac{1}{2}|z|^2 + \frac{1-\mu}{r} + \frac{\mu}{r'} \qquad r^2 = (x+\mu)^2 + y^2,$$

$$r'^2 = (x-1+\mu)^2 + y^2.$$

When $r = 0$ or $r' = 0$ $(r, r'$ are distances), the equation appears to be collision singular points. But if we use new variables $w = u + iv$, $\tau$ such as

$$z = \frac{1}{2} \cos w$$

$$d = \left|\frac{dz}{dw}\right|^2 dt = rr'dt.$$

(1) can be transformed as

$$\frac{d^2 w}{d\tau^2} + 2irr' \frac{dw}{d\tau} = \text{grad}_w \bar{\Omega} \tag{13}$$

where

$$\bar{\Omega} = rr'(\Omega + h),$$

h is a constant. The equation (13) is analytic in the whole w-plane.

Yi, et al. [6,7] use the regularized equation to study some qualitative properties of the configuration manifolds and their tangent space of this dynamical system by means of numerical methods

BIBLIOGRAPHY

[1] Soubbotin, M. F., Improvement of the convergence of the basic development on theory of perturbative motion. Bull. Inst. Theor. Astron. (Leningrad), Vol 4, No. 1, 1947, p. 54-64.

[2] Yi, Zhao-Hua, Improvement of the convergence of the development of disturbing function (Planar Case). Acta Astronomica Sinica, Vol. 5, No. 1, 1957, pp. 61-78.

[3] Yi, Zhao-Hua, Ibid. vol. 5, No.2, 1957, pp. 265-275.

[4] He, Miao-Fu, Improvement of the convergence of the development of disturbing function of spatial three-body problem. Acta Astronomica Sinica. Vol. 6, No. 2, 1958, pp. 129-229.

[5] Dong, Min-De, New exact method for Hill's equation. Acta Astronomica Sinica. Vol. 21, No. 1, 1980, pp. 51-57.

[6] Jefferys, W. H. and Yi, Zhao-Hua, On the stability of restricted three-body problem by means of LCN. Celestial Mechanics. Vol. 30, No. 1, 1983, pp. 85-95

[7] Yi, Zhao-Hua and Jefferys, W. H., On the computing method of LCN of restricted three-body problem. Acta Astronomica Sinica, Vol. 24, No. 1, 1983, pp. 58-67.

Department of Astronomy
Nanjing University
Nanjing

Contemporary Mathematics
Volume 48, 1985

# COMPLEX VARIABLE METHODS IN ELASTICITY

Jian-ke Lu (Chien-ke Lu)

It is well known that the complex variable method is very effective in plane elasticity. N.I. Muskhelishvili's monograph "Some Basic Problems of the Mathematical Theory of Elasticity" (Groningen, Noordhoff, 1963) is most famous in this direction. He actually was the father of Russian school in this branch. In China, research works in this field began in the early 60's of this century and developed rapidly in the last ten years. A brief sketch about the scientific works in this area in China will be given in this paper. Emphasis is put on the mathematical theory.

1. Basic notions and terminology. Let $S$ be a multiply connected elastic region in the complex plan. Assume in general the material is isotropic. As usual, $\sigma_x, \sigma_y, \tau_{xy}$ denote the stresses and $u + iv$ denotes the displacement. They are functions of $z \in S$. Two complex Airy function $\phi(z)$ and $\psi(z)$ are connected with them by the equalities

$$\sigma_x + \sigma_y = 4\mathrm{Re}\,\phi'(z), \tag{1}$$

$$\sigma_y - \sigma_x + 2i\tau_{xy} = 2\{\bar{z}\phi''(z)+\psi'(z)\}, \tag{2}$$

$$2\mu(u+iv) = \kappa\phi(z) - z\overline{\phi'(z)} - \overline{\psi(z)}, \tag{3}$$

where $\kappa, \mu$ are elastic constants relying on the medium. $\phi(z)$ and $\psi(z)$ are analytic but multi-valued in general in $S$ and continuous to its boundary $L = \partial S$. However, $\phi'(z) = \Phi(z)$ and $\psi'(z) = \Psi(z)$ are single-valued, i.e., holomorphic in $S$. The external stress applied to $L$ at the point $t$ is denoted by $X_n(t) + iY_n(t)$ ($n$ being the external normal of $L$ at $t$). The positive sense of $L$ will be chosen such that $S$ is situated on the left of $L$. Hence

$$X + iY = \int_L [X_n(t)+iY_n(t)]ds$$


0271-4132/85 $1.00 + $.25 per page

(s is the arc-length parameter of t) is the principal vector and

$$\mathrm{Re}\int_L f(t)d\bar{t}$$

is the principal moment of the external forces applied on L, where

$$f(t) = i\int_{t_0}^{t} [X_n(t)+iY_n(t)]ds.$$

There are two basic problems in elasto-statics.

The first fundamental problem (I). Given the external stresses on L, find the elastic equilibrium.

The second fundamental problem (II). Given the displacements on L, find the elastic equilibrium.

By means of (1) - (3), these problems are transferred to certain boundary value problems of analytic functions, namely,

$$k\phi(t) + t\overline{\phi'(t)} + \overline{\psi(t)} = f^*(t), \quad t \in L, \tag{4}$$

where $f^*(t)$ is a given function on L, while $k = 1$ for problem (I) and $k = -\kappa$ for problem (II). After suitable treatment, the unknown functions $\phi(z)$ and $\psi(z)$ are already holomorphic in S.

Many authors solved these boundary value problems by reducing them to Fredholm integral equations along L by means of Cauchy-type integral representations of $\phi(z)$ and $\psi(z)$. The most eminent method, as author's opinion, was due to D.I. Sherman (Cf. Muskhelishvili's monograph mentioned above, §102).

2. Problems related to fundamental ones. In 1964, the author studied the general fundamental problems with cyclic symmetry [9], i.e., L is cyclic with respect to the origin as well as the boundary conditions applied on it. By conformal mapping, Sherman's equations are simplified in the sense that the path of integration is reduced to a non-congruent part of it. The method was proven to be effective in some special cases by Rong Er-qian [1].

For the case when S is the infinite plane with some circular holes, Tang Li-min [1] gave its stress analysis by expanding Airy functions into Laurent series and using the method of undetermined coefficients.

When S is the infinite plane with some arbitrary holes, for the problem (II), Muskhelishvili had proved that, besides the displacements on L and the conditions at infinity, it must also be given the resultant principal vector of external stresses along L so as to obtain the unique solution of the problem. The author posed new formulations of the problem (II) in [10]: the displacement on each boundary contour of $L$ is given only relative to a rigid motion which may be different to each other and there are also given the principal vector and the principal moment of the external stresses applied to each boundary contour. It was shown that the solution is also unique in this case, either for $S$ infinite or bounded.

The author also investigated the welding problems of different materials [3], when there appear displacement differences between interfaces. The number and shape of the different media is quite arbitrary. The boundary value conditions now become that, besides the condition (4), in which $k$ may be different for different boundary contours of $L$ for problem (II), we must have, on the interface $\Gamma_{ij}$ of the subregions $S_i$ and $S_j$ of different media,

$$\phi^+(\tau) + \tau\overline{\phi'^+(\tau)} + \overline{\psi^+(\tau)} = \phi^-(\tau) + \tau\overline{\phi'^-(\tau)} + \overline{\psi^-(\tau)}, \quad \tau \in \Gamma_{ij}$$

(the equilibrium of stresses) and

$$\alpha_i\phi^+(\tau) + \beta_i[\tau\overline{\phi'^+(\tau)}+\overline{\psi^+(\tau)}]$$

$$= \alpha_j \phi^-(\tau) + \beta_j[\tau\overline{\phi'^-(\tau)}+\overline{\psi^-(\tau)}] + f_{ij}(\tau) \qquad \tau \in \Gamma_{ij},$$

where $\alpha_i$, $\beta_i$ and $\alpha_j$, $\beta_j$ are known constants determined by the elastic constants of $S_i$ and $S_j$ respectively, and $f_{ij}(\tau)$ is also a given function. Such problems were reduced to singular integral equations (SIE) along the interfaces and the boundary $L$, by using the methods modified to Sherman's. Analytic solutions were obtained for the case when $S$ is the whole plane and the interface is the real axis or a circle.

3. Crack problems. It is of great importance in theory of fracture for the case when the boundary of $S$ contain cracks. Now the condition (4) is replaced by the analogous equalities for $t$ belonging to the different sides of each crack and $\phi(z)$, $\psi(z)$ are allowed to have singularities of integral type at the tips of cracks.

An infinite plane with single line-segment crack is classical. An infinite plane with cracks arbitrary in number and shape (assuming smooth) was considered by the author in [2] and then the same problem for bounded region in [11]. All these problems were reduced to SIE again by methods modified to Sherman's.

Special crack problems were studied by various authors. An infinite plane with a broken line-segment crack was discussed by Wang Zi-qiang [1] by using conformal mapping. The problem of a ring-shaped region with a single crack was solved effectively by Tang Rem-ji and Wang Kai [3], which was also reduced to SIE. The problem of the infinite plane with an elliptic hole weakened by a system of line-segment cracks was solved by Tang and Wang Yin-bang [1], in which the numerical method of solution of the obtained SIE is illustrated.

The Mellin transform is another useful tool for some special problems, which was used frequently by F. Erdogan in U.S. It was used to solve plane elastic problems with radial cracks by Tang and Wang Kai [2], and Tang and Jiang Zhu-zhong [1,2]. The method is also used for torsion problems, e.g., Tang [1,2], Tang and Wang Kai [1], Guo Zhong-heng [1,2].

The method of series expansion (Guo [3], Wang Keren [1], Ouyang Chang and Zhou Shiao-kang [1]) and the method of conformal mapping (Wang Kai-fu [1], Ouyang [1], Zhou Cheng-fan and Guan Chang-wen [1]) were also used to solve some special problems.

The crack problems of compound materials bonded by two half planes were discussed by F. Erdogan in many papers when cracks are straight segments and radial, by using Mellin transform. The author considered the case when cracks are curvilinear and arbitrary [8]. Such problems were reduced to SIE along the interface and the cracks. They were then further simplified to SIE along the cracks only, which could much lighten the calculations. The method was shown to be applicable to the case when the interface is a circle [9] and even if in general case [12].

Many results are collected in the author's monograph [13].

4. Periodic problems. If the boundary $L$ of $S$ consists of periodic contours (or cracks) and the boundary conditions subjected to $L$ are the same for congruent points (mod $p$, the period), then it is called a periodic problem. There are problems (I)' and (II)' corresponding to (I) and (II) respectively. The infinite plane with a row of periodic circular holes was considered in literature.

The author obtained many results relating to periodic problems (including circular welding, periodic stamps applied to the boundary of a half-plane, etc.) in [1] by reducing them to Riemann boundary value problems of analytic functions.

In [5,7], the author proved that the stresses are periodic iff the displacements are quasi-periodic, i.e.,

$$u(z+p) + iv(z+p) = u(z) + iv(z) + q, \qquad z \in S,$$

where $q$ is certain constant. Thus, the proper formulations of such problems were clarified.

Periodic contact problems were studied by Cai Hai-tao [1,2]. The periodic problems of infinite plane with arbitrary cracks were discussed by the author in [6] and then generalized by Cai to the anisotropic case [4,5]. The problems (I)' and (II)' for the half-plane with anisotropic elasticity were also discussed by Cai [3]. He transferred these problems to Hilbert boundary value problems of analytic functions.

The periodic problem (I)' of an elastic strip was studied by Liu Shi-qiang [1]. The periodic crack problem of bonded half-planes of different media was studied by Ma Dao-wei [1].

There has been published a monograph about periodic problems by the author and Cai [1].

The doubly periodic problems were rarely discussed. Now we have only the paper written by Hao Tiao-fu, Cheng Ji-da and Yu Jia-sheng [1]. Using conformal mapping, they discussed the doubly periodic crack problem under certain special boundary conditions.

5. Miscellaneous problems. Dynamic problems of orthogonal an isotropic half-plane loaded by moving pressure were discussed by Sun Kuo-ying [1].

The problem of moving crack with constant speed in the infinte plane was discussed by Cai [6].

A special problem of multi-valued displacement was studied by Tang Ren-ji and Zhen Ji-qing [1].

## REFERENCES

All are written in Chinese (usually with English abstract) except those marked with asterisk.

Cai Hai-tao

1. Problem of the periodic contact of elastic theory for the isotropic plane, J. of Central-South Inst. Mining and Metallurgy, 1979, No. 1, 113-125.

2. The problem of periodic contact in the plane theory of elasticity, Acta Math. Appl. Sinica, 2 (1979), 181-195.

3. On the first and the second periodic fundamental problems of semi-infintie medium with anisotropic elasticity, Acts Mech. Sinica, 11 (1979), 240-247.

4. Periodic crack problem of plane anisotropic medium, Acta Math. Sci., 2 (1982), 35-44.

5. The periodic cracks of an infinite anisotropic medium for plane skew-symmetric loadings (to appear in Acta Mech. Solida Sinica).

6. Movable crack problem (to appear in Acta Math. Sci.).

Guo Zhong-heng

1. The stress intensity factor of twisted circular cylinder with radiating internal cracks, Acta Mech. Sinica, 11 (1979), 298-302.

2. On the torsion of circular cylinder with a single arbitrary radial crack, Acta Mech. Sinica, 12 (1980), 208-212.

3. Bending of concentrically cracked circular cylinder and tube, Acta Mech. Sinica, 12 (1980), 423-427.

Liu Shi-qiang

1. On periodic fundamental problems of an elastic strip, J. of Math. (PRC), 4 (1984), 165-176.

Lu Jian-ke

1. Periodic Riemann boundary value problem and its application to elasticity, Acta Math. Sinica, 13 (1963), 343-388.

2. On fundamental problems for the infinite elastic plane with cracks, J. Wuhan Univ. (math. special ed.), 1963, No. 2, 37-49.

3. On the plane welding problems of different materials, J. Wuhan Univ. (math. special ed.), 1963, No. 2, 50-66.

4. On mathematical problems of elastic plane with cyclic symmetry, J. Wuhan Univ., 1964, No. 2, 1-13.

5. On fundamental problems of plane elasticity with periodic stresses, Acta Mech. Sinica, 7 (1964), 316-327.

6. On problems of an infinte elastic plane with arbitrary periodic cracks, J. of Central-South Inst. of Mining and Metallurgy, 1980, No. 2, 9-19.

7. A remark on plane elasticity with periodic stresses, J. Wuhan Univ., 1980, No. 2, 9-10.

8.* Circular welding problems with a crack, Appl. Math. Mech. (english ed.), 4 (1983), 751-763.

9. New formulations of the second fundamental problem in plane elasticity, Appl. Math. Mech., 6 (1985), 223-230.

10. The mathematical problems of bounded region with cracks (to appear in J. Wuhan Univ.).

11. The mathematical problems of compound materials with cracks in plane elasticity (to appear in Acts Math. Sci.).

12. Complex Variable Methods in Plane Elasticity (to be published by Wuhan Un. Press).

Lu Jian-ke, Cai Hai-tao

1. Periodic Plane Elastic Problems, Hunan People's Press of Sci. and Tech., Changsha, 1985.

Ma Dao-wei

1. The periodic crack problems of compound materials (to appear).

Ouyang Chang

1. On a class of methods for solving problems of random boundary notches and/or cracks, Appl, Math. Mech., 1 (1980), 159-166.

Ouyang Chang, Zhou Shiao-kang

1. One computations of stress intensity factors for stiffened half-planes with imperfections, Appl. Math. Mech., 6 (1985), 121-133.

Rong Ěr-qian

1. Some plane elastic problems on the circular symmetric multiply connected region, Acta Math. Appl. Sinica, 5 (1982), 388-396.

Sun Kuo-ying

1. The boundary problems of dynamic plane orthogonal anisotropic elasticity, Acta Mech. Sinica, 5 (1962), 254-259.

Tan Li-min

1. Stress analysis of several adjacent circular holes in elastic plane, Sci. Record, 10 (1959), 366-375.

Tang Ren-ji

1. Saint-Venant's torsion problem for a circular cylinder with cracks, Acts Mech. Sinica, 14 (1982), 332-346.

2. Bending torsion of circular cylinder with radial crack system, Acta Mech. Solida Sinica, 1983, No. 3, 341-353.

Tang Ren-ji, Jiang Zhu-zheng
1. An analysis of radial crack system with a centric circular hole, J. of Lanzhou Univ., 1979, 226-247.

Tang Ren-ji, Wang Kai
1. Bending of a thin cracked plate, J. of Lanzhou Univ., 1979, 192-201.

2.* On the Griffith crack whose surfaces are loaded asymmetrically, Eng. Fracture Mech., 16 (1982), 47-54.

3. An elasticity solution of a single crack problem in plane ring-shaped region, Acta Mech. Sinica, 14 (1982), 546-557.

Tang Ren-ji, Wang Yin-bang
1. Problem of crack system with an elliptic hole, Acta Mech. Sinica, 16 (1984), 570-579.

Tang Ren-ji, Zhen Ji-qing
1.* General solution of multi-valued displacement problem of an eccentric circular ring, Appl. Math. Mech. (English ed.), 4 (1983), 743-750.

Wang Kai-fu
1. The torsion problem of a circular cylinder with circular arc cracks, Acta Mech. Sinica, 5 (1962), 58-62.

Wang Ke-ren
1. The weighting functions in fracture mechanics, their generalization and calculations, Acta Mech. Sinica, 13 (1981), 569-580.

Wang Zi-qiang
1. Fracture criteria for cracks of compound type, Reprint of Inst. of Mech., Chinese Acad. of Sci., 1977.

Zhou Cheng-fan, Guan Cheng-wen
1. Stress concentration and stress intensity factors for an infinite plane with several rows of elliptic holes and cracks. Appl. Math. Mech., 4 (1983), 789-800.

Department of Mathematics
Wuhan University, Wuhan

ABCDEFGHIJ–AMS–898765